AS-Level Physics

AS Physics is seriously tricky — no question about that.
To do well, you're going to need to revise properly and practise hard.

This book has thorough notes on all the theory you need,
and it's got practice questions... lots of them.
For every topic there are warm-up and exam-style questions.

And of course, we've done our best to make the whole thing vaguely entertaining for you.

Complete Revision and Practice

Exam Board: OCR A

Editors:
Amy Boutal, Julie Wakeling, Sarah Williams

Contributors
Tony Alldridge, Jane Cartwright, Peter Cecil, Peter Clarke, Mark A. Edwards, Barbara Mascetti, John Myers, Andy Williams

Proofreaders:
Ian Francis, Glenn Rogers

Published by CGP

Data used to construct stopping distance diagram on page 26 from the Highway Code. Reproduced under the terms of the Click-Use Licence.

With thanks to Jan Greenway for the copyright research.

ISBN: 978 1 84762 131 3

Groovy website: www.cgpbooks.co.uk
Jolly bits of clipart from CorelDRAW®
Printed by Elanders Ltd, Newcastle upon Tyne.

Based on the classic CGP style created by Richard Parsons.

Contents

How Science Works

The Scientific Process 2

Unit 1: Section 1 — Motion

Scalars and Vectors 4
Motion with Constant Acceleration 6
Free Fall 8
Free Fall and Projectile Motion 10
Displacement-Time Graphs 12
Velocity-Time Graphs 14

Unit 1: Section 2 — Forces in Action

Newton's Laws of Motion 16
Drag and Terminal Velocity 18
Mass, Weight and Centre of Gravity 20
Forces and Equilibrium 22
Moments and Torques 24
Car Safety 26

Unit 1: Section 3 — Work and Energy

Work and Power 28
Conservation of Energy 30
Efficiency and Sankey Diagrams 32
Hooke's Law 34
Stress and Strain 36
The Young Modulus 38
Interpreting Stress-Strain Curves 40

Unit 2: Section 1 — Electric Current, Resistance and DC Circuits

Charge, Current and Potential Difference 42
Resistance and Resistivity 44
I/V Characteristics 46
Electrical Energy and Power 48
Domestic Energy and Fuses 50
E.m.f. and Internal Resistance 52
Conservation of Energy and Charge in Circuits .. 54
The Potential Divider 56

Unit 2: Section 2 — Waves

The Nature of Waves 58
Longitudinal and Transverse Waves 60
The Electromagnetic Spectrum 62
Superposition and Coherence 64
Standing (Stationary) Waves 66
Diffraction 68
Two-Source Interference 70
Diffraction Gratings 72

Unit 2: Section 3 — Quantum Phenomena

Light — Wave or Particle 74
The Photoelectric Effect 76
Energy Levels and Photon Emission 78
Wave-Particle Duality 80

Answering Experiment Questions

Error Analysis 82

Answers 84

Index 92

How Science Works is all about the scientific process — how we develop and test scientific ideas. It's what scientists do all day, every day (well, except at coffee time — never come between a scientist and their coffee).

Scientists Come Up with **Theories** — Then **Test Them**...

Science tries to explain **how** and **why** things happen — it **answers questions**. It's all about seeking and gaining **knowledge** about the world around us. Scientists do this by **asking** questions and **suggesting** answers and then **testing** them, to see if they're correct — this is the **scientific process**.

1) **Ask** a question — make an **observation** and ask **why or how** it happens. E.g. what is the nature of light?
2) **Suggest** an answer, or part of an answer, by forming:
    - a **theory** (a possible **explanation** of the observations) e.g. light is a wave.
    - a **model** (a **simplified picture** of what's physically going on)
3) Make a **prediction** or **hypothesis** — a **specific testable statement**, based on the theory, about what will happen in a test situation. E.g. light should interfere and diffract.
4) Carry out a **test** — to provide **evidence** that will support the prediction, or help disprove it. E.g. Young's double-slit experiment.

The evidence supported Quentin's Theory of Flammable Burps.

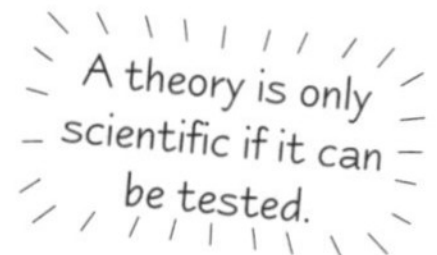

A theory is only scientific if it can be tested.

...Then They **Tell** Everyone About Their **Results**...

The results are **published** — scientists need to let others know about their work. Scientists publish their results in **scientific journals**. These are just like normal magazines, only they contain **scientific reports** (called papers) instead of the latest celebrity gossip.

1) Scientific reports are similar to the **lab write-ups** you do in school. And just as a lab write-up is **reviewed** (marked) by your teacher, reports in scientific journals undergo **peer review** before they're published.
2) The report is sent out to **peers** — other scientists that are experts in the **same area**. They examine the data and results, and if they think that the conclusion is reasonable it's **published**. This makes sure that work published in scientific journals is of a **good standard**.
3) But peer review **can't guarantee** the science is **correct** — other scientists still need to **reproduce** it.
4) Sometimes **mistakes** are made and bad work is published. Peer review **isn't perfect** but it's probably the best way for scientists to self-regulate their work and to publish **quality reports**.

...Then **Other Scientists** Will **Test** the Theory Too

Other scientists read the published theories and results, and try to **test the theory** themselves. This involves:

- Repeating the **exact same experiments**.
- Using the theory to make **new predictions** and then testing them with **new experiments**.

If the **Evidence** Supports a Theory, It's **Accepted — for Now**

1) If all the experiments in all the world provide evidence to back it up, the theory is thought of as **scientific 'fact'** (for now).
2) But they never become **totally undisputable** fact. Scientific **breakthroughs or advances** could provide new ways to question and test the theory, which could lead to **new evidence** that **conflicts** with the current evidence. Then the testing starts all over again...

And this, my friend, is the **tentative nature of scientific knowledge** — it's always **changing** and **evolving**.

The Scientific Process

So scientists need evidence to back up their theories. They get it by carrying out experiments, and when that's not possible they carry out studies. But why bother with science at all? We want to know as much as possible so we can use it to try and improve our lives (and because we're nosey).

Evidence Comes From Controlled Lab Experiments...

1) Results from **controlled experiments** in **laboratories** are **great**.
2) A lab is the easiest place to **control variables** so that they're all **kept constant** (except for the one you're investigating).

For example, finding the resistance of a piece of material by altering the voltage across the material and measuring the current flowing through it (see p. 46). All other variables need to be kept the same, e.g. the dimensions of the piece of material being tested, as they may also affect its resistance.

... That You can Draw Meaningful Conclusions From

1) You always need to make your experiments as **controlled** as possible so you can be confident that any effects you see are linked to the variable you're changing.
2) If you do find a relationship, you need to be careful what you conclude. You need to decide whether the effect you're seeing is **caused** by changing a variable, or whether the two are just **correlated**.

"Right Geoff, you can start the experiment now... I've stopped time..."

Society Makes Decisions Based on Scientific Evidence

1) Lots of scientific work eventually leads to **important discoveries** or breakthroughs that could **benefit humankind**.
2) These results are **used by society** (that's you, me and everyone else) to **make decisions** — about the way we live, what we eat, what we drive, etc.
3) All sections of society use scientific evidence to make decisions, e.g. politicians use it to devise policies and individuals use science to make decisions about their own lives.

Other factors can **influence** decisions about science or the way science is used:

Economic factors

- Society has to consider the **cost** of implementing changes based on scientific conclusions — e.g. the cost of reducing the UK's carbon emissions to limit the human contribution to **global warming**.
- Scientific research is often **expensive**. E.g. in areas such as astronomy, the Government has to **justify** spending money on a new telescope rather than pumping money into, say, the **NHS** or **schools**.

Social factors

- **Decisions** affect **people's lives** — e.g. when looking for a site to build a **nuclear power station**, you need to consider how it would affect the lives of the people in the **surrounding area**.

Environmental factors

- Many scientists suggest that building **wind farms** would be a **cheap** and **environmentally friendly** way to generate electricity in the future. But some people think that because **wind turbines** can **harm wildlife** such as birds and bats, other methods of generating electricity should be used.

So there you have it — how science works...

Hopefully these pages have given you a nice intro to how science works, e.g. what scientists do to provide you with 'facts'. You need to understand this, as you're expected to know how science works yourself — for the exam and for life.

Scalars and Vectors

Mechanics is one of those things that you either love or hate. I won't tell you which side of the fence I'm on.

Scalars Only Have Size, but Vectors Have Size and Direction

1) A **scalar** has **no direction** — it's **just an amount** of something, like the **mass** of a **sack of meaty dog food.**
2) A **vector** has magnitude (**size**) and **direction** — like the **speed and direction** of next door's **cat** running away.
3) **Force** and **velocity** are both **vectors** — you need to know **which way** they're going as well as **how big** they are.
4) Here are a few examples to get you started:

Scalars	Vectors
mass, temperature, time, length, speed, energy	displacement, force, velocity, acceleration, momentum

Adding Vectors Involves Pythagoras and Trigonometry

Adding two or more vectors is called finding the **resultant** of them.
You find the resultant of two vectors by drawing them **'tip-to-tail'**.

Example

Jemima goes for a walk. She walks 3 m North and 4 m East. She has walked 7 m but she isn't 7 m from her starting point. Find the magnitude and direction of her displacement.

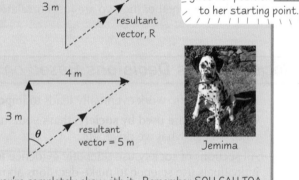

Jemima's 'displacement' gives her position relative to her starting point.

First, draw the vectors **tip-to-tail**. Then draw a line from the **tail** of the first vector to the **tip** of the last vector to give the **resultant**:
Because the vectors are at right angles, you get the **magnitude** of the resultant using Pythagoras:

4 m
3 m
resultant vector, R

$R^2 = 3^2 + 4^2 = 25$
So $R = 5$ m

Now find the **bearing** of Jemima's new position from her original position.
You use the triangle again, but this time you need to use trigonometry. You know the opposite and the adjacent sides, so you need to use:

4 m
3 m
θ
resultant vector = 5 m

Jemima

$\tan\theta = 4/3$

$\theta = 53.1°$ Trig's really useful in mechanics — so make sure you're completely okay with it. Remember SOH CAH TOA.

Use the Same Method for Resultant Forces or Velocities

If the vectors aren't at right angles, you'll need to do a scale drawing.

Always start by drawing a diagram.

Example

2 N
add vectors tip-to-tail
2 N
R
2 N
2 N

You know the resultant force is at 45° to the horizontal (since both forces are the same size).
So all you need to do is use Pythagoras:

$R^2 = 2^2 + 2^2 = 8$

which gives $R = 2.83$ N at 45° to the horizontal.

Don't forget to take the square root.

Example

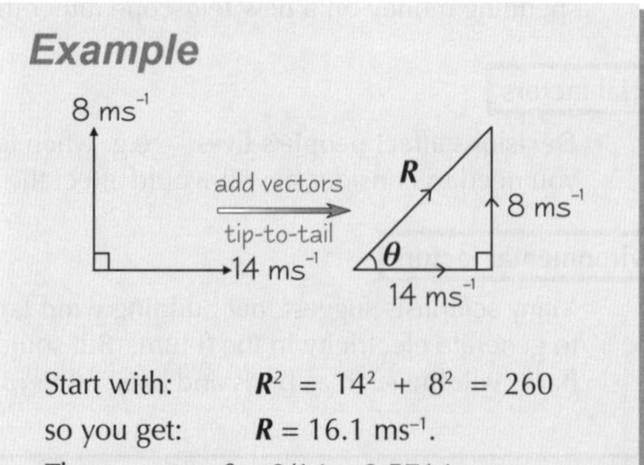

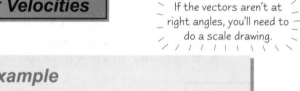

Start with: $R^2 = 14^2 + 8^2 = 260$

so you get: $R = 16.1$ ms^{-1}.

Then: $\tan\theta = 8/14 = 0.5714$

$\theta = 29.7°$

Scalars and Vectors

Sometimes you have to do it backwards.

*It's Useful to Split a **Vector** into **Horizontal** and **Vertical Components***

This is the opposite of finding the resultant — you start from the resultant vector and split it into two **components** at right angles to each other. You're basically **working backwards** from the examples on the other page.

Resolving a vector v into horizontal and vertical components

See pages 22 and 23 for more on resolving.

You get the **horizontal** component v_x like this:

$$\cos\theta = v_x / v$$

$$v_x = v\cos\theta$$

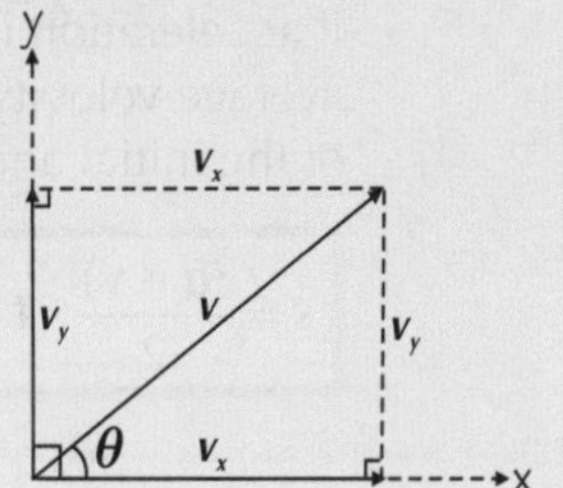

...and the **vertical** component v_y like this:

$$\sin\theta = v_y / v$$

$$v_y = v\sin\theta$$

θ is measured anticlockwise from the horizontal.

Example

Charley's amazing floating home is travelling at a speed of 5 ms^{-1} at an angle of 60° up from the horizontal. Find the vertical and horizontal components.

Charley's mobile home was the envy of all his friends.

The **horizontal** component v_x is:

$v_x = v\cos\theta = \mathbf{5\cos 60°} = 2.5$ ms^{-1}

The vertical component v_y is:

$v_y = v\sin\theta = \mathbf{5\sin 60°} = 4.33$ ms^{-1}

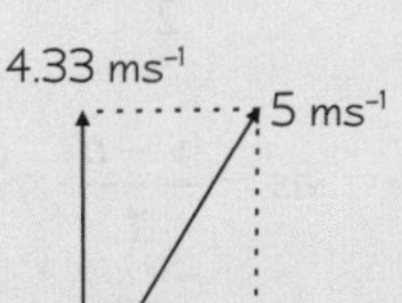

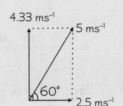

Resolving is dead useful because the two components of a vector **don't affect each other**. This means you can deal with the two directions **completely separately**.

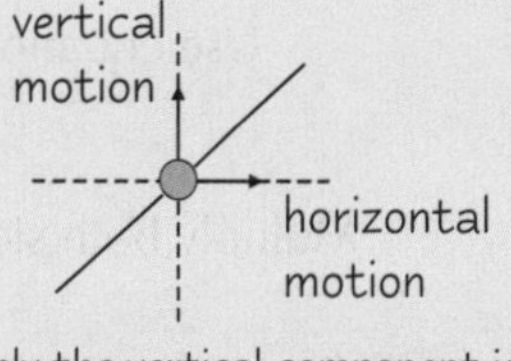

Only the vertical component is affected by gravity.

Practice Questions

Q1 Explain the difference between a scalar quantity and a vector quantity.

Q2 Jemima has gone for a swim in a river which is flowing at 0.35 ms^{-1}. She swims at 0.18 ms^{-1} at right angles to the current. Show that her resultant velocity is 0.39 ms^{-1} at an angle of 27.2° to the current.

Q3 Jemima is pulling on her lead with a force of 40 N at an angle of 26° below the horizontal. Show that the horizontal component of this force is about 36 N.

Exam Questions

Q1 The wind is creating a horizontal force of 20 N on a falling rock of weight 75 N. Calculate the magnitude and direction of the resultant force. [2 marks]

Q2 A glider is travelling at a velocity of 20.0 ms^{-1} at an angle of 15° below the horizontal. Find the horizontal and vertical components of the glider's velocity. [2 marks]

His Dark Vectors Trilogy — displacement, velocity and acceleration...

Well there's nothing like starting the book on a high. And this is nothing like... yes, OK. Ahem. Well, good evening folks. I'll mostly be handing out useful information in boxes like this. But I thought I'd not rush into it, so this one's totally useless.

Uniform Acceleration is Constant Acceleration

Acceleration could mean a change in speed or direction or both.

Uniform means **constant** here. It's nothing to do with what you wear.

There are **four main equations** that you use to solve problems involving **uniform acceleration**.
You need to be able to use them, **and** how they're **derived**.

1) **Acceleration is the rate of change of velocity.**
From this definition you get:

$$a = \frac{(v-u)}{t} \quad \text{so} \quad \boxed{v = u + at}$$

where:
u = initial velocity, a = acceleration
v = final velocity, t = time taken

2) **s = average velocity × time**
If acceleration is constant, the average velocity is just the average of the initial and final velocities, so:

$$\boxed{s = \frac{(u+v)}{2} \times t} \quad s = \text{displacement}$$

3) Substitute the expression for v from equation 1 into equation 2 to give:

$$s = \frac{(u+u+at)\times t}{2} = \frac{2ut+at^2}{2} \qquad \boxed{s = ut + \tfrac{1}{2}at^2}$$

4) You can **derive** the fourth equation from equations **1** and **2**:

Use equation **1** in the form: $a = \frac{v-u}{t}$

Multiply both sides by s, where: $s = \frac{(u+v)}{2} \times t$

This gives us: $as = \frac{(v-u)}{t} \times \frac{(u+v)t}{2}$

The t's on the right cancel, so:
$2as = (v-u)(v+u)$
$2as = v^2 - uv + uv - u^2$

so: $\boxed{v^2 = u^2 + 2as}$

Example

A tile falls from a roof 25 m high. Calculate its speed when it hits the ground and how long it takes to fall. Take $g = 9.8\ \text{ms}^{-2}$.

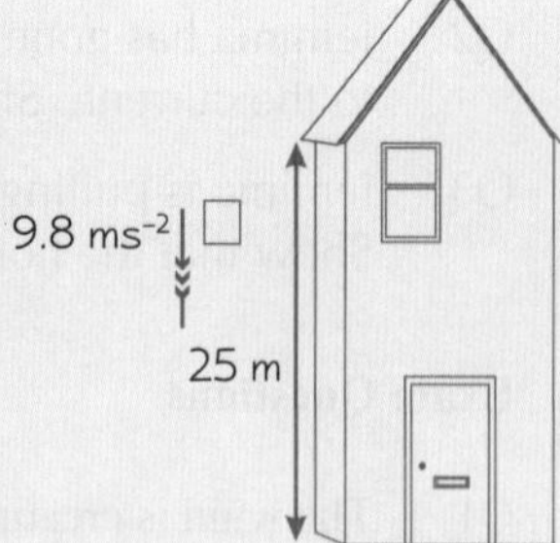

First of all, write out what you know:
$s = 25$ m
$u = 0\ \text{ms}^{-1}$ since the tile's stationary to start with
$a = 9.8\ \text{ms}^{-2}$ due to gravity
$v = ?$ $t = ?$

Usually you take upwards as the positive direction. In this question it's probably easier to take downwards as positive, so you get $g = +9.8\ \text{ms}^{-2}$ instead of $g = -9.8\ \text{ms}^{-2}$.

Then, choose an equation with only **one unknown quantity**.
So start with $v^2 = u^2 + 2as$
$v^2 = 0 + 2 \times 9.8 \times 25$
$v^2 = 490$
$v = 22.1\ \text{ms}^{-1}$

Now, find t using:
$s = ut + \tfrac{1}{2}at^2$
$25 = 0 + \tfrac{1}{2} \times 9.8 \times t^2$
$t^2 = \frac{25}{4.9}$

Final answers:
$t = 2.3$ s
$v = 22.1\ \text{ms}^{-1}$

Motion with Constant Acceleration

Example

A car accelerates steadily from rest at a rate of 4.2 ms^{-2} for 6 seconds.

a) Calculate the final speed.

b) Calculate the distance travelled in 6 seconds.

4.2 ms^{-2}

Remember — always start by writing down what you know.

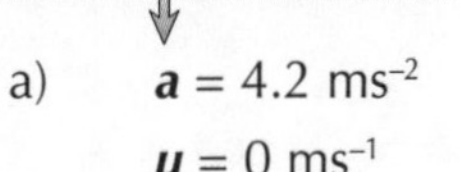

a)
$a = 4.2\ ms^{-2}$
$u = 0\ ms^{-1}$
$t = 6\ s$
$v = ?$

choose the right equation... $v = u + at$

$v = 0 + 4.2 \times 6$

Final answer: $v = 25.2\ ms^{-1}$

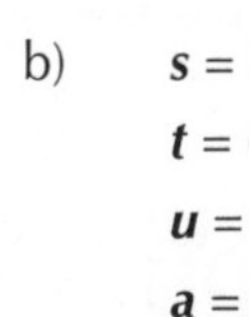

b)
$s = ?$
$t = 6\ s$
$u = 0\ ms^{-1}$
$a = 4.2\ ms^{-2}$
$v = 25.2\ ms^{-1}$

you can use: $s = \frac{(u+v)t}{2}$ or: $s = ut + \frac{1}{2}at^2$

$s = \frac{(0+25.2)\times 6}{2}$

$s = 0 + \frac{1}{2} \times 4.2 \times (6)^2$

Final answer: $s = 75.6$ m

$s = 75.6$ m

*You Have to **Learn** the Constant Acceleration **Equations***

Make sure you learn the equations. There are only four of them and these questions are always dead easy marks in the exam, so you'd be dafter than a hedgehog in a helicopter not to learn them...

Practice Questions

Q1 Write out the four constant acceleration equations.

Q2 Show how the equation $s = ut + \frac{1}{2}at^2$ can be derived.

Exam Questions

Q1 A skydiver jumps from an aeroplane when it is flying horizontally. She accelerates due to gravity for 5 s.
(a) Calculate her maximum vertical velocity. (Assume no air resistance.) [2 marks]
(b) How far does she fall in this time? [2 marks]

Q2 A motorcyclist slows down uniformly as he approaches a red light. He takes 3.2 seconds to come to a halt and travels 40 m in this time.
(a) How fast was he travelling initially? [2 marks]
(b) Calculate his acceleration. (N.B. a negative value shows a deceleration.) [2 marks]

Q3 A stream provides a constant acceleration of 6 ms^{-2}. A toy boat is pushed directly against the current and then released from a point 1.2 m upstream from a small waterfall. Just before it reaches the waterfall, it is travelling at a speed of 5 ms^{-1}.
(a) Find the initial velocity of the boat. [2 marks]
(b) What is the maximum distance upstream from the waterfall the boat reaches? [2 marks]

Constant acceleration — it'll end in tears...

If a question talks about "uniform" or "constant" acceleration, it's a dead giveaway they want you to use one of these equations. The tricky bit is working out which one to use — start every question by writing out what you know and what you need to know. That makes it much easier to see which equation you need. To be sure. Arrr.

Free Fall

So, how do you work this parachute thing agaiAAAAAaaaaaarrrrrggghhhhhhhhhhhhhh...

Free Fall is when there's Only Gravity and Nothing Else

Free fall is defined as "the motion of an object undergoing an acceleration of '*g*'".
You need to remember:

1) Acceleration is a **vector quantity** — and '*g*' acts **vertically downwards**.
2) Unless you're given a different value, take the magnitude of *g* as **9.81 ms^{-2}**, though it varies slightly at different points on the Earth's surface.
3) The **only force** acting on an object in free fall is its **weight**.
4) Objects can have an initial velocity in any direction and still undergo **free fall** as long as the **force** providing the initial velocity is **no longer acting**.

You Can Measure g by using an Object in Free Fall

You don't have to do it this way — but if you don't know a method of measuring *g* already, learn this one.

You need to be able to:

1) **Sketch** a diagram of the **apparatus.**
2) **Describe** the **method.**
3) **List** the **measurements** you make.
4) **Explain** how '*g*' is **calculated.**
5) Be aware of sources of **error.**

Another gravity experiment.

Experiment to Measure the Acceleration Due to Gravity

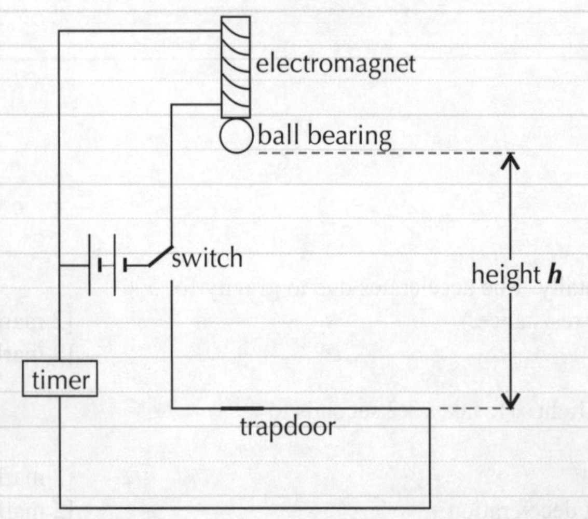

In this experiment you have to assume that the effect of air resistance on the ball bearing is negligible and that the magnetism of the electromagnet decays instantly.

The Method:

1) Measure the height *h* from the **bottom** of the ball bearing to the **trapdoor**.
2) Flick the switch to simultaneously start the timer and disconnect the electromagnet, releasing the ball bearing.
3) The ball bearing falls, knocking the trapdoor down and breaking the circuit — which stops the timer.

Use the time *t* measured by the timer, and the height *h* that the ball bearing has fallen, to calculate a value for **g**, using $h = \frac{1}{2}gt^2$ (see next page for more on acceleration formulas).

The most significant source of error in this experiment will be in the measurement of *h*.
Using a ruler, you'll have an uncertainty of about 1 mm.
This dwarfs any error from switch delay or air resistance.

Free Fall

You can Just Replace a with g in the Equations of Motion

You need to be able to work out **speeds**, **distances** and **times** for objects in **free fall**. Since g is a **constant acceleration** you can use the **constant acceleration equations**. But g acts downwards, so you need to be careful about directions. To make it clear, there's a sign convention: **upwards is positive**, **downwards is negative**.

Sign Conventions — Learn Them:

g is always downwards so it's usually negative	t is always positive
u and v can be either positive or negative	s can be either positive or negative

Case 1: No initial velocity (it just falls)

Initial velocity $u = 0$

Acceleration $a = g = -9.81\ \text{ms}^{-2}$

So the constant acceleration equations become: ⟹

$$v = gt \qquad v^2 = 2gs$$
$$s = \frac{1}{2}gt^2 \qquad s = \frac{vt}{2}$$

Case 2: An initial velocity upwards (it's thrown up into the air)

The constant acceleration equations are just as normal, but with $a = g = -9.81\ \text{ms}^{-2}$

Case 3: An initial velocity downwards (it's thrown down)

Example: Alex throws a stone down a cliff. She gives it a downwards velocity of $2\ \text{ms}^{-1}$. It takes 3 s to reach the water below. How high is the cliff?

1) You know $u = -2\ \text{ms}^{-1}$, $a = g = -9.81\ \text{ms}^{-2}$ and $t = 3$ s. You need to find s.

s will be negative because the stone ends up further down than it started

2) Use $s = ut + \frac{1}{2}gt^2 = (-2\times3) + \left(\frac{1}{2}\times -9.81\times 3^2\right) = -50.1$ m. **The cliff is 50.1 m high.**

Practice Questions

Q1 What is the value of the acceleration of a free-falling object?

Q2 What is the initial velocity of an object which is dropped?

Exam Questions

Q1 A student has designed a device to estimate the value of 'g'. It consists of two narrow strips of card joined by a piece of transparent plastic. The student drops the device through a light gate connected to a computer. As the device falls, the strips of card break the light beam.

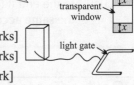

(a) What three pieces of data will the student need from the computer to estimate 'g'? [3 marks]

(b) Explain how the measurements from the light gate can be used to estimate 'g'. [3 marks]

(c) Give one reason why the student's value of 'g' will not be entirely accurate. [1 mark]

card
transparent window
x
x
light gate

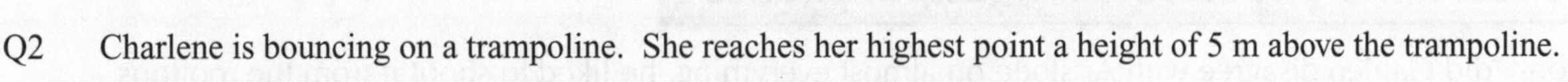

Q2 Charlene is bouncing on a trampoline. She reaches her highest point a height of 5 m above the trampoline.

(a) Calculate the speed with which she leaves the trampoline surface. [2 marks]

(b) How long does it take her to reach the highest point? [2 marks]

(c) What will her velocity be as she lands back on the trampoline? [2 marks]

It's not the falling that hurts — it's the being pelted with rotten vegetables... okay, okay...

The hardest bit with free fall questions is getting your signs right. Draw yourself a little diagram before you start doing any calculations, and label it with what you know and what you want to know. That can help you get the signs straight in your head. It also helps the person marking your paper if it's clear what your sign convention is. Always good.

Free Fall and Projectile Motion

What goes up, must come down — but no one really questioned why until Aristotle. He thought he knew... but then Galileo and Newton sure showed him...

Aristotle — Heavy Objects Fall Quicker than Lighter objects

1) **Aristotle** was an ancient Greek philosopher who sat around thinking about pretty much everything, including the **joys of Physics**.
2) He used **reasoning** to try and work out how the world worked from **everyday** observations.
3) One of his famous theories was that if **two objects** of **different mass** are dropped from the **same height**, the **heavier** object would always hit the ground **before** the lighter object.

Trev was counting... there was no way Tez took longer to hit the ground than him.

Galileo — All Objects in Free Fall Accelerate Uniformly

1) Galileo thought that **all objects accelerate towards the ground at the same rate** — so objects with different weights dropped from the same height should hit the ground at the **same time**.
2) Not only that, but he reckoned the reason objects didn't seem to do this was because of the effect of **air resistance** on different objects.
3) Believe it or not, scientists don't think Galileo chucked stuff from the top of the Leaning Tower of Pisa to test this theory. Instead he did an even more exciting experiment — he.. er... rolled balls down a slope.

Example: The Inclined Plane Experiment

1) Handy things like stop clocks and light gates hadn't been invented, so Galileo had to find a way of slowing down the free fall of an object without otherwise affecting to have any chance of showing it accelerating.
2) Ta da — the inclined plane experiment was born. Galileo found that by rolling a ball down a **smooth** groove in an inclined plane, he **reduced** the effect of **air resistance** while slowing the ball's fall at the same time.
3) He timed how long it took the ball to roll from the top of the slope to the bottom using a water clock.
4) By rolling the ball along different fractions of the total length of the slope, he found that the distance the ball travelled was proportional to the square of the time taken. The ball was **accelerating** at a **constant rate**.
5) In the end it took Newton to bring it all together and explain why **all** free-falling objects have the **same acceleration** (see page 17).

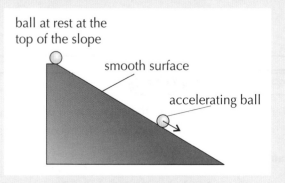

ball at rest at the top of the slope

smooth surface

accelerating ball

Galileo Tested his Theories using Experiments

1) Not only did Galileo disagree with Aristotle on almost everything, he liked to shout it from the rooftops. He not only managed to insult other philosophers at the time, but the Pope and the entire Catholic church too — which got him in a whole **load of trouble**.
2) Even though he was so unpopular, Galileo's theories eventually overturned Aristotle's and became **generally accepted**. He wasn't the first person to question Aristotle, but his success was down to the **systematic** and **rigorous experiments** he used to **test** his theories. These experiments could be repeated and the results described **mathematically** and compared.
3) It's not much different in science now. You make a prediction and test it — the more **scientific evidence** that supports your theory, the more **accepted** it becomes.

Free Fall and Projectile Motion

Any object given an initial velocity and then left to move freely under gravity is a projectile.
If you're doing AS Maths, you've got all this to look forward to in M1 as well, quite likely. Fun for all the family.

*You have to think of **Horizontal** and **Vertical** Motion **Separately***

Example

Sharon fires a scale model of a TV talent show presenter horizontally with a velocity of 100 ms^{-1} from 1.5 m above the ground. How long does it take to hit the ground, and how far does it travel? Assume the model acts as a particle, the ground is horizontal and there is no air resistance.

Think about vertical motion first:

1) It's **constant acceleration** under gravity...
2) You know $\boldsymbol{u} = 0$ (no vertical velocity at first), $\mathbf{s} = -1.5$ m and $\mathbf{a} = \mathbf{g} = -9.81\ ms^{-2}$. You need to find $\boldsymbol{t}$.

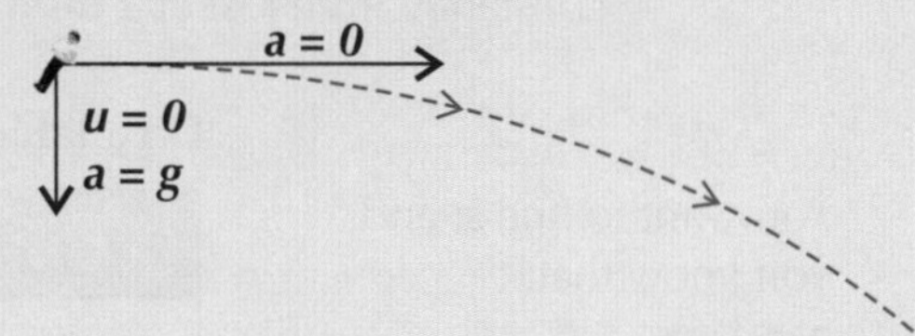

3) Use $s = \frac{1}{2}gt^2 \Rightarrow t = \sqrt{\frac{2s}{g}} = \sqrt{\frac{2\times -1.5}{-9.81}} = 0.55$ s

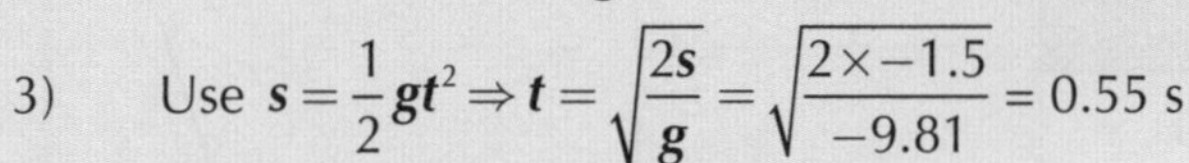

4) So the model hits the ground after **0.55** seconds.

Then do the horizontal motion:

1) The horizontal motion isn't affected by gravity or any other force, so it moves at a **constant speed**.
2) That means you can just use good old **speed = distance / time**.
3) Now $v_h = 100\ ms^{-1}$, $t = 0.55$ s and $a = 0$. You need to find s_h.
4) $s_h = v_h t = 100 \times 0.55 =$ **<u>55 m</u>**

Where v_h is the horizontal velocity, and s_h is the horizontal distance travelled (rather than the height fallen).

*It's **Slightly Trickier** if it **Starts Off** at an **Angle***

If something's projected at an angle (like, say, a javelin) you start off with both horizontal and vertical velocity:

<u>Method</u>:
1) Resolve the initial velocity into horizontal and vertical components.
2) Use the vertical component to work out how long it's in the air and/or how high it goes.
3) Use the horizontal component to work out how far it goes while it's in the air.

Practice Questions

Q1 What is the initial vertical velocity for an object projected horizontally with a velocity of 5 ms^{-1}?

Q2 How does the horizontal velocity of a free-falling object change with time?

Q3 What is the main reason Galileo's ideas became generally accepted in place of Aristotle's?

Exam Questions

Q1 Jason stands on a vertical cliff edge throwing stones into the sea below.
He throws a stone horizontally with a velocity of 20 ms^{-1}, 560 m above sea level.
(a) How long does it take for the stone to hit the water from leaving Jason's hand?
Use g = 9.81 ms^{-2} and ignore air resistance. [2 marks]
(b) Find the distance of the stone from the base of the cliff when it hits the water. [2 marks]

Q2 Robin fires an arrow into the air with a vertical velocity of 30 ms^{-1}, and a horizontal velocity of 20 ms^{-1}, from 1 m above the ground. Find the maximum height from the ground reached by his arrow.
Use g = 9.81 ms^{-2} and ignore air resistance. [3 marks]

<u>So it's this "Galileo" geezer who's to blame for my practicals...</u>

Ah, the ups and downs and er... acrosses of life. Make sure you're happy splitting an object's motion into horizontal and vertical bits — it comes up all over mechanics. Hmmm... I wonder what Galileo would be proudest of, insisting on the systematic, rigorous experimental method on which modern science hangs... or getting in a Queen song? Magnificoooooo...

Drawing graphs by hand — oh joy. You'd think examiners had never heard of the graphical calculator. Ah well, until they manage to drag themselves out of the dark ages, you'll just have to grit your teeth and get on with it.

Acceleration Means a **Curved Displacement-Time Graph**

A graph of displacement against time for an **accelerating object** always produces a **curve**.
If the object is accelerating at a **uniform rate**, then the **rate of change** of the **gradient** will be constant.

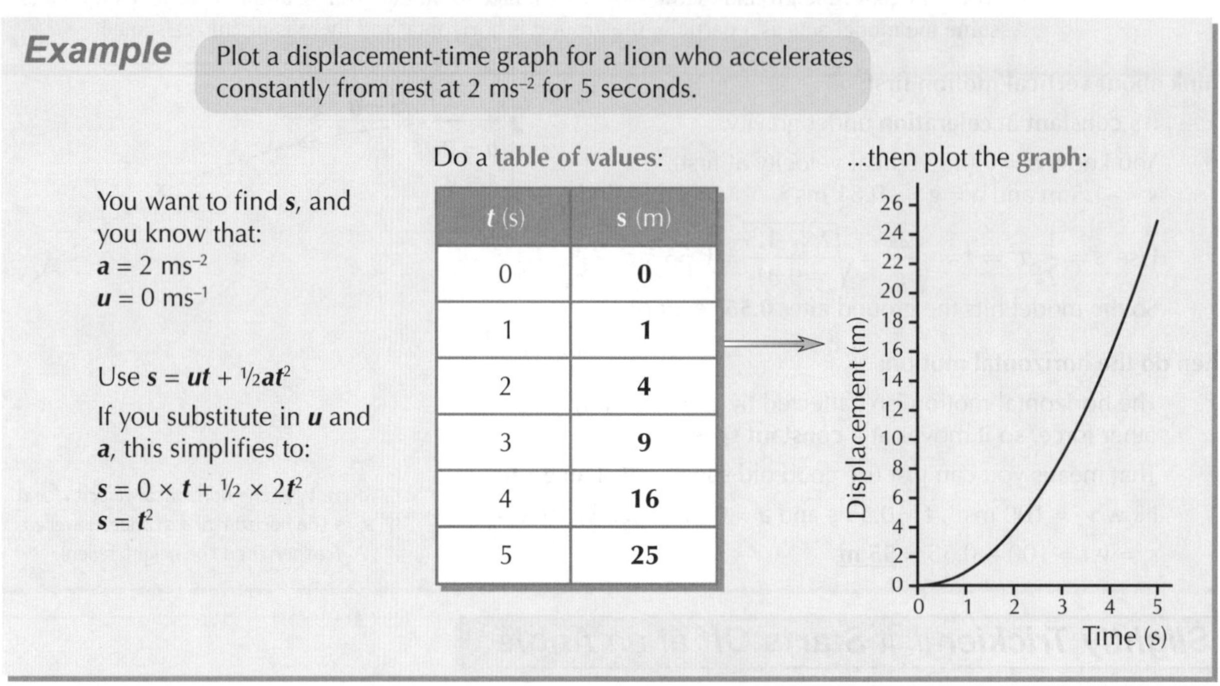

Example Plot a displacement-time graph for a lion who accelerates constantly from rest at 2 ms^{-2} for 5 seconds.

You want to find **s**, and you know that:
$a = 2\ ms^{-2}$
$u = 0\ ms^{-1}$

Use $s = ut + \frac{1}{2}at^2$

If you substitute in ***u*** and ***a***, this simplifies to:

$s = 0 \times t + \frac{1}{2} \times 2t^2$
$s = t^2$

Do a **table of values:**

t (s)	s (m)
0	**0**
1	**1**
2	**4**
3	**9**
4	**16**
5	**25**

...then plot the **graph:**

Different Accelerations Have **Different Gradients**

In the example above, if the lion has a **different acceleration** it'll change the **gradient** of the curve like this:

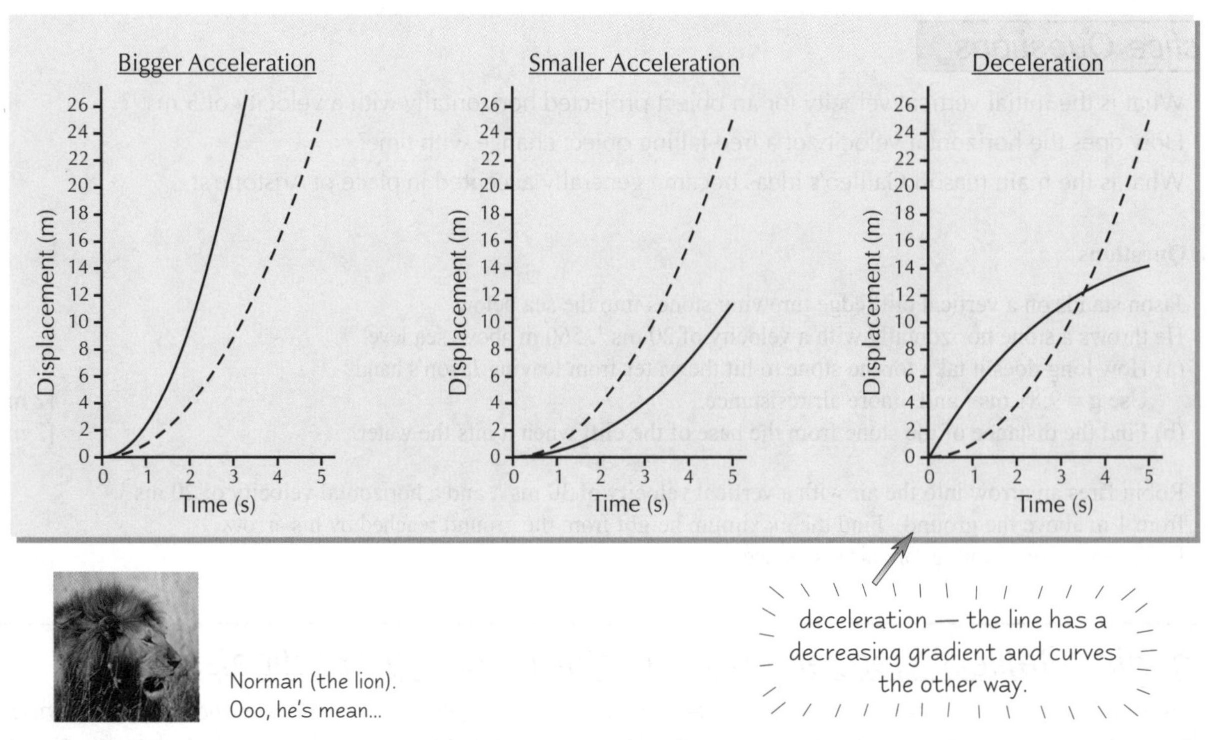

Norman (the lion).
Ooo, he's mean...

deceleration — the line has a decreasing gradient and curves the other way.

Displacement-Time Graphs

The **Gradient** of a **Displacement-Time Graph** Tells You the Velocity

When the velocity is constant, the graph's a **straight line**.
Velocity is defined as...

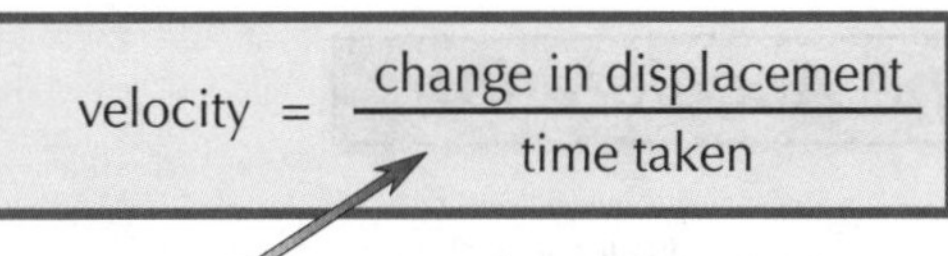

On the graph, this is $\frac{\text{change in } y\ (\Delta y)}{\text{change in } x\ (\Delta x)}$, i.e. the gradient.

So to get the velocity from a displacement-time graph, just find the gradient.

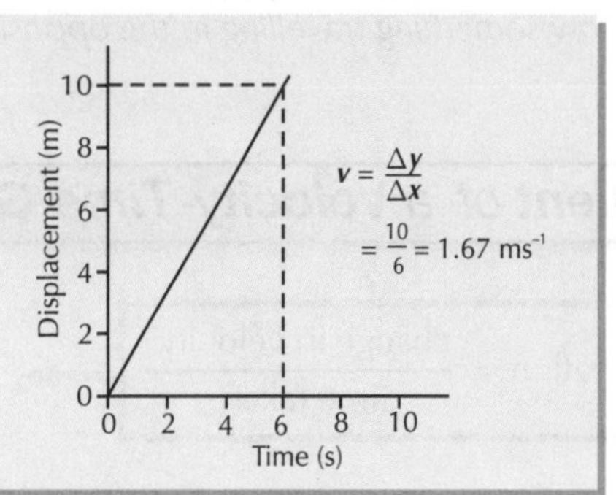

It's the Same with **Curved Graphs**

If the gradient **isn't constant** (i.e. if it's a curved line), it means the object is **accelerating**.

> To find the **velocity** at a certain point you need to draw a **tangent** to the curve at that point and find its gradient.

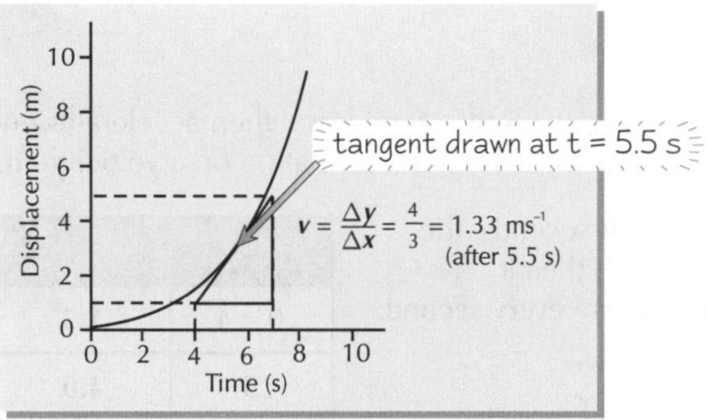

Practice Questions

Q1 What is given by the slope of a displacement-time graph?

Q2 Sketch a displacement-time graph to show: a) constant velocity, b) acceleration, c) deceleration

Exam Questions

Q1 Describe the motion of the cyclist as shown by the graph below. [4 marks]

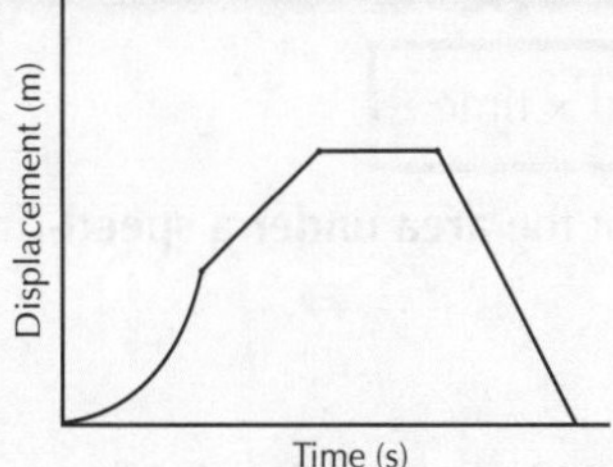

Q2 A baby crawls 5 m in 8 seconds at a constant velocity. She then rests for 5 seconds before crawling a further 3 m in 5 seconds. Finally, she makes her way back to her starting point in 10 seconds, travelling at a constant speed all the way.

(a) Draw a displacement-time graph to show the baby's journey. [4 marks]

(b) Calculate her velocity at all the different stages of her journey. [2 marks]

Some curves are bigger than others...

Whether it's a straight line or a curve, the steeper it is, the greater the velocity. There's nothing difficult about these graphs — the main problem is that it's easy to get them muddled up with velocity-time graphs (next page). If in doubt, think about the gradient — is it velocity or acceleration, is it changing (curve), is it constant (straight line), is it 0 (horizontal line)...

Velocity-Time Graphs

Speed-time graphs and velocity-time graphs are pretty similar. The big difference is that velocity-time graphs can have a negative part to show something travelling in the opposite direction:

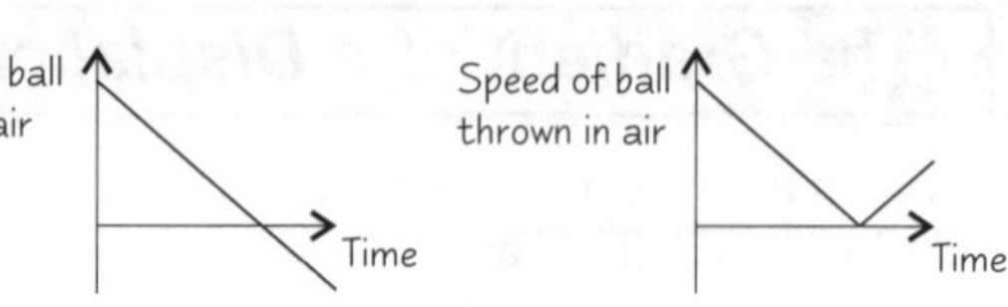

The **Gradient** of a **Velocity-Time Graph** tells you the **Acceleration**

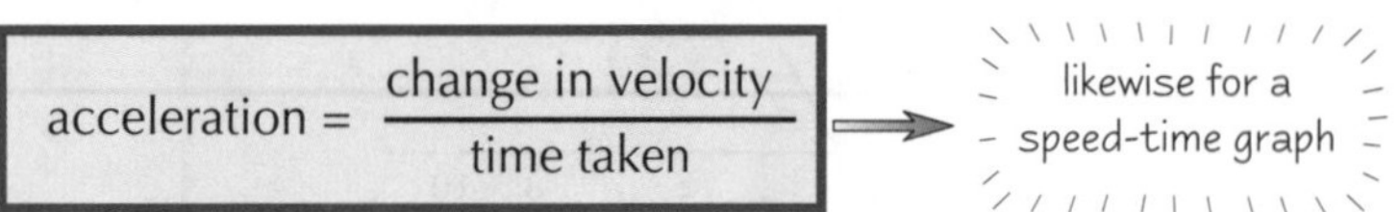

$$\text{acceleration} = \frac{\text{change in velocity}}{\text{time taken}}$$

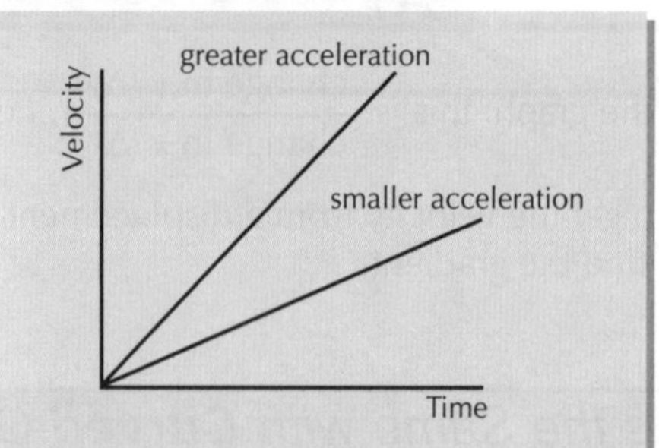

So the acceleration is just the **gradient** of a **velocity-time graph**.

1) **Uniform** acceleration is always a **straight line**.
2) The **steeper** the **gradient**, the **greater** the **acceleration**.

Example

A lion strolls along at 1.5 ms^{-1} for 4 s and then accelerates uniformly at a rate of 2.5 ms^{-2} for 4 s. Plot this information on a velocity-time graph.

So, for the first four seconds, the velocity is 1.5 ms^{-1}, then it increases by **2.5 ms^{-1} every second**:

t (s)	v (ms^{-1})
0 – 4	**1.5**
5	**4.0**
6	**6.5**
7	**9.0**
8	**11.5**

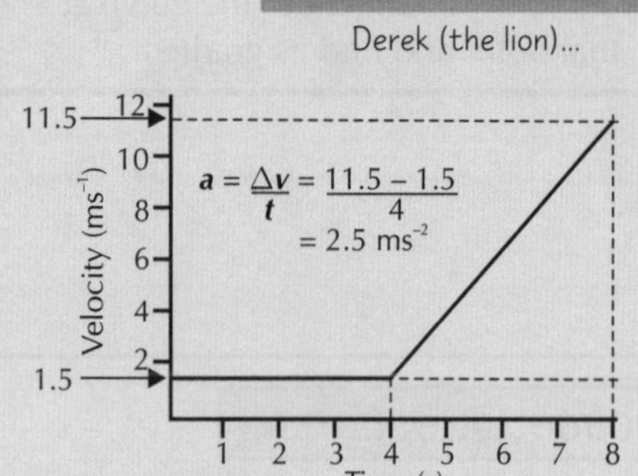

Derek (the lion)...

11.5, 1.5, Velocity (ms^{-1}), Time (s)

$$a = \frac{\Delta v}{t} = \frac{11.5 - 1.5}{4} = 2.5\ ms^{-2}$$

You can see that the **gradient of the line** is **constant** between 4 s and 8 s and has a value of 2.5 ms^{-2}, representing the **acceleration of the lion**.

Distance Travelled = **Area** under **Speed-Time Graph**

You know that:

distance travelled = average speed × time

So you can find the distance travelled by working out the **area under a speed-time graph**.

Example

A racing car accelerates uniformly from rest to 40 ms^{-1} in 10 s. It maintains this speed for a further 20 s before coming to rest by decelerating at a constant rate over the next 15 s. Draw a velocity-time graph for this journey and use it to calculate the total distance travelled by the racing car.

Split the **graph** up into **sections:** A, B and C

Calculate the **area** of each and **add** the three results together.

A: Area = ½ base × height = ½ × 10 × 40 = 200 m

B: Area = b × h = 20 × 40 = 800 m

C: Area = ½ b × h = ½ × 15 × 40 = 300 m

Total distance travelled = 1300 m

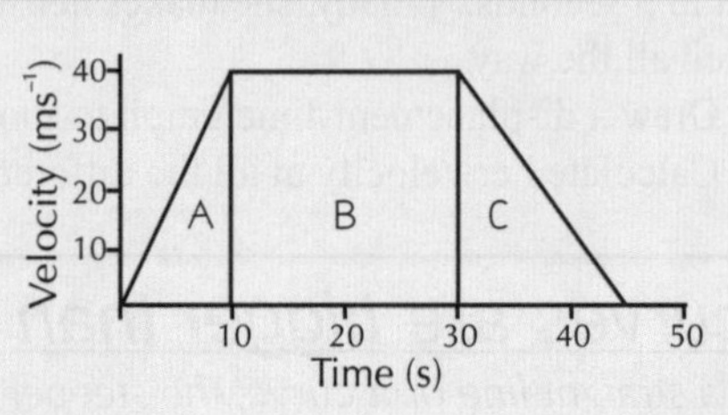

Velocity-Time Graphs

Non-Uniform Acceleration is a Curve on a V-T Graph

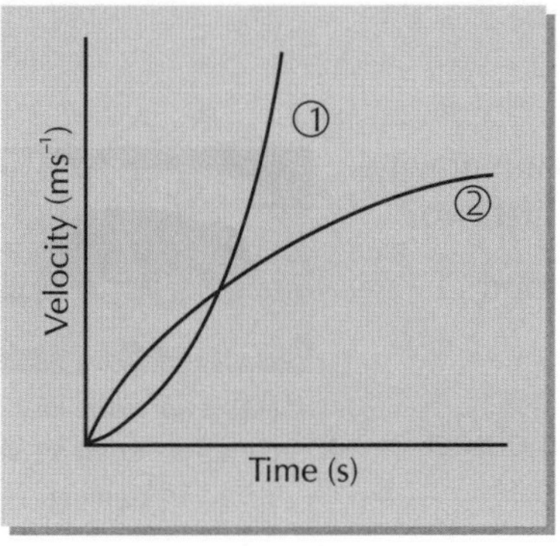

1) If the acceleration is changing, the gradient of the velocity-time graph will also be changing — so you **won't** get a **straight line**.
2) **Increasing acceleration** is shown by an **increasing gradient** — like in curve ①.
3) **Decreasing acceleration** is shown by a **decreasing gradient** — like in curve ②.

Simple enough...

You Can Draw Displacement-Time and Velocity-Time Graphs Using ICT

Instead of gathering distance and time data using **traditional methods**, e.g. a stopwatch and ruler, you can be a bit more **high-tech**.

A fairly **standard** piece of kit you can use for motion experiments is an **ultrasound position detector**. This is a type of **data-logger** that automatically records the **distance** of an object from the sensor several times a second.

If you attach one of these detectors to a computer with **graph-drawing software**, you can get **real-time** displacement-time and velocity-time graphs.

The main **advantages** of data-loggers over traditional methods are:

1) The data is more **accurate** — you don't have to allow for human reaction times.
2) Automatic systems have a much higher **sampling** rate than humans — most ultrasound position detectors can take a reading ten times every second.
3) You can see the data displayed in **real time**.

Practice Questions

Q1 How do you calculate acceleration from a velocity-time graph?

Q2 How do you calculate the distance travelled from a speed-time graph?

Q3 Sketch velocity-time graphs for constant velocity and constant acceleration.

Q4 Describe the main advantages of ICT over traditional methods for the collection and display of motion data.

Exam Question

Q1 A skier accelerates uniformly from rest at 2 ms^{-2} down a straight slope.

(a) Sketch a velocity-time graph for the first 5 s of his journey. [2 marks]

(b) Use a constant acceleration equation to calculate his displacement at t = 1, 2, 3, 4 and 5 s, and plot this information onto a displacement-time graph. [5 marks]

(c) Suggest another method of calculating the skier's distance travelled after each second and use this to check your answers to part (b). [2 marks]

Still awake — I'll give you five more minutes...

There's a really nice sunset outside my window. It's one of those ones that makes the whole landscape go pinky-yellowish. And that's about as much interest as I can muster on this topic. Normal service will be resumed on page 16, I hope.

You did most of this at GCSE, but that doesn't mean you can just skip over it now. You'll be kicking yourself if you forget this stuff in the exam — easy marks...

Newton's Laws are Only Approximations

Isaac Newton's a pretty famous chap. Not only was he inspired by an apple to explain gravity, he also wrote three really important scientific laws — you need to know about the first two.

1) Newton's laws work pretty well. At **everyday speeds** they give really, really good approximations — but they're **not** the whole story.
2) At very **high speeds**, you have to take into account **relativistic effects**. According to the **Special Theory of Relativity**, as you increase the speed of an object its **mass** increases. So mass isn't constant, and $F = ma$ doesn't work any more.

Gerty swore she lost 10 pounds as soon as she stopped running.

Newton's 1st Law says that a Force is Needed to Change Velocity

1) **Newton's 1st law of motion** states that the **velocity** of an object will **not change** unless a **net force** acts on it.
2) In plain English this means a body will stay still or move in a **straight line** at a **constant speed**, unless there's a **net force** acting on it.
3) If the forces **aren't balanced**, the **overall net force** will make the body **accelerate**. This could be a change in **direction**, or **speed**, or both. (See Newton's 2nd law, below.)

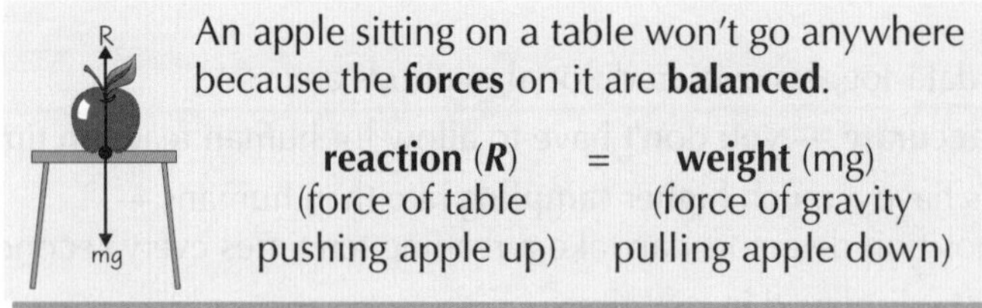

An apple sitting on a table won't go anywhere because the **forces** on it are **balanced**.

reaction (***R***) = **weight** (mg)
(force of table pushing apple up) (force of gravity pulling apple down)

Newton's 2nd Law says that Acceleration is Proportional to the Force

...which can be written as the well-known equation:

net force (N) = mass (kg) × acceleration (ms^{-2})

$$F = m \times a$$

Learn this — it crops up all over the place in AS Physics. And learn what it means too:

1) It says that the **more force** you have acting on a certain mass, the **more acceleration** you get.
2) It says that for a given force the **more mass** you have, the **less acceleration** you get.
3) Galileo said that all objects fall at the same rate if you ignore air resistance — and that man wasn't wrong. You can see **why** with a bit of ball dropping and a dash of Newton's 2nd law — check out the next page.

REMEMBER:
1) The **net force** is the **vector sum** of all the forces.
2) The force is **always** measured in **newtons**.
3) The **mass** is always measured in **kilograms**.
4) The **acceleration** is always in the **same direction** as the **net force**.

Newton's Laws of Motion

Acceleration is Independent of Mass

Imagine dropping two balls at the same time — ball **1** being heavy, and ball **2** being light.
Then use Newton's 2nd law to find their acceleration.

mass = m_1
resultant force = F_1
acceleration = a_1
By Newton's second law:

$$F_1 = m_1a_1$$

Ignoring air resistance, the only force acting on the ball is weight, given by $W_1 = m_1g$ (where g = gravitational field strength = 9.81 Nkg^{-1}).

So: $F_1 = m_1a_1 = W_1 = m_1g$

So: $m_1a_1 = m_1g$, then m_1 cancels out to give: $a_1 = g$

mass = m_2
resultant force = F_2
acceleration = a_2
By Newton's second law:

$$F_2 = m_2a_2$$

Ignoring air resistance, the only force acting on the ball is weight, given by $W_2 = m_2g$ (where g = gravitational field strength = 9.81 Nkg^{-1}).

So: $F_2 = m_2a_2 = W_2 = m_2g$

So: $m_2a_2 = m_2g$, then m_2 cancels out to give: $a_2 = g$

... in other words, the **acceleration** is **independent of the mass**. It makes **no difference** whether the ball is **heavy or light**. And I've kindly **hammered home the point** by showing you two almost identical examples.

Practice Questions

Q1 State Newton's 1st and 2nd laws of motion, and explain what they mean.

Q2 Ball A has a mass of 5 kg and ball B has a mass of 3 kg.
Both balls are dropped from the same height at the same time — which one hits the ground first? Why?

Exam Questions

Q1 Draw diagrams to show the forces acting on a parachutist:
(i) accelerating downwards. [1 mark]
(ii) having reached terminal velocity. [1 mark]

Q2 A boat is moving across a river. The engines provide a force of 500 N at right angles to the flow of the river and the boat experiences a drag of 100 N in the opposite direction. The force on the boat due to the flow of the river is 300 N. The mass of the boat is 250 kg.
(a) Calculate the magnitude of the net force acting on the boat. [2 marks]
(b) Calculate the magnitude of the acceleration of the boat. [2 marks]

Q3 This question asks you to use Newton's second law to explain three situations.
(a) Two cars have different maximum accelerations.
What are the only two overall factors that determine the acceleration a car can have? [2 marks]
(b) Michael can always beat his younger brother Tom in a sprint, however short the distance.
Give two possible reasons for this. [2 marks]
(c) Michael and Tom are both keen on diving. They notice that they seem to take the same time to drop from the diving board to the water. Explain why this is the case. [3 marks]

Newton's 4th Law... avoid the brain-eating apples falling from the sky...

These laws may not really fill you with a huge amount of excitement (and I hardly blame you if they don't)... but it was pretty fantastic at the time — suddenly people actually understood how forces work, and how they affect motion. I mean arguably it was one of the most important scientific discoveries ever...

If you jump out of a plane at 2000 ft, you want to know that you're not going to be accelerating all the way.

Friction is a Force that Opposes Motion

There are two main types of friction:

1) **Contact friction** between **solid surfaces** (which is what we usually mean when we just use the word 'friction'). You don't need to worry about that too much for now.
2) **Fluid friction** (known as **drag** or fluid resistance or air resistance).

Fluid Friction or Drag:

1) 'Fluid' is a word that means either a **liquid or a gas** — something that can **flow**.
2) The force depends on the thickness (or **viscosity**) of the fluid.
3) It **increases** as the **speed increases** (for simple situations it's directly proportional, but you don't need to worry about the mathematical relationship).
4) It also depends on the **shape** of the object moving through it — the larger the **area** pushing against the fluid, the greater the resistance force.

Things you need to remember about frictional forces:

1) They **always** act in the **opposite direction** to the **motion** of the object.
2) They can **never** speed things up or start something moving.
3) They convert **kinetic energy** into **heat**.

Terminal Velocity — when the Friction Force Equals the Driving Force

You will reach a **terminal velocity** at some point, if you have:

1) a **driving force** that stays the **same** all the time
2) a **frictional** or **drag force** (or collection of forces) that increases with speed

There are **three main stages** to reaching terminal velocity:

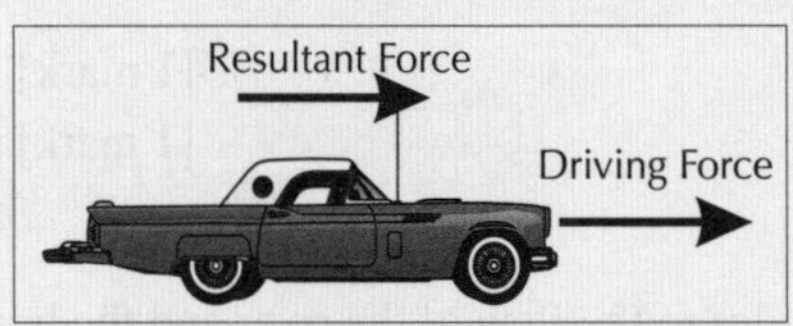

The car **accelerates** from **rest** using a constant driving force.

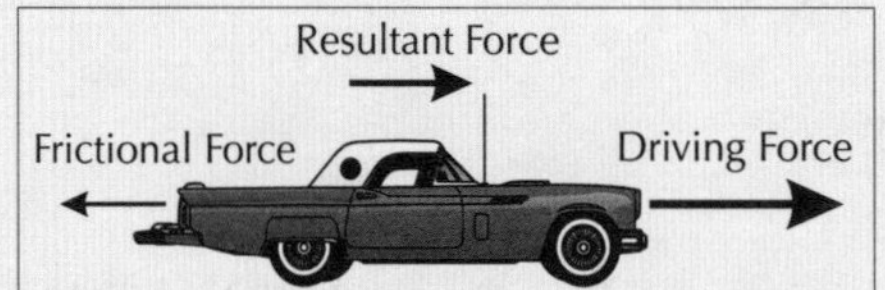

As the **velocity increases**, the **resistance forces increase** (because of things like turbulence — you don't need the details). This **reduces the resultant force** on the car and hence **reduces its acceleration**.

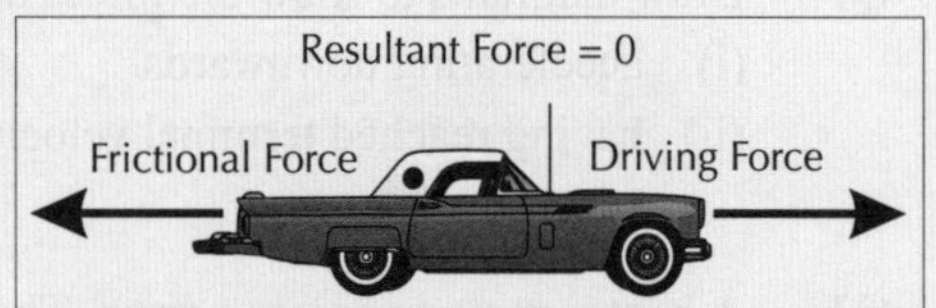

Eventually the car reaches a velocity at which the **resistance forces are equal to the driving force**. There is now **no resultant force** and **no acceleration**, so the car carries on at **constant velocity**.

Sketching a Graph for Terminal Velocity

You need to be able to **recognise** and **sketch** the graphs for **velocity against time** and **acceleration against time** for the **terminal velocity** situation.

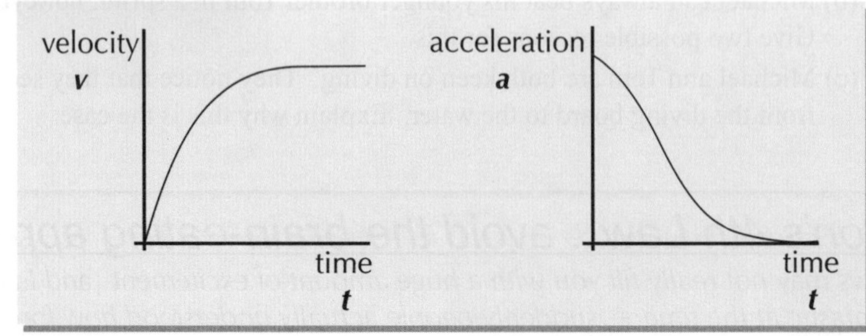

Nothing for it but practice — shut the book and sketch them from memory. Keep doing it till you get them right every time.

Drag and Terminal Velocity

*Things **Falling** through **Air** or **Water** Reach a **Terminal Velocity** too*

When something's falling through **air**, the **weight** of the object is a **constant force** accelerating the object downwards. **Air resistance** is a **frictional force** opposing this motion, which **increases** with **speed**.
So before a parachutist opens the parachute, exactly the same thing happens as with the car example:

1) A skydiver leaves a plane and will **accelerate** until the **air resistance** equals his **weight**.

2) He will then be travelling at a **terminal velocity**.

But... the terminal velocity of a person in free fall is too great to land without dying a horrible death.
The **parachute increases** the **air resistance massively**, which slows him down to a lower terminal velocity:

3) Before reaching the ground he will **open his parachute**, which immediately **increases the air resistance** so it is now **bigger** than his **weight**.

4) This **slows him down** until his speed has dropped enough for the **air resistance** to be **equal to his weight** again. This new terminal velocity is small enough to survive landing.

The v-t graph is a bit different, because you have a new terminal velocity being reached after the parachute is opened:

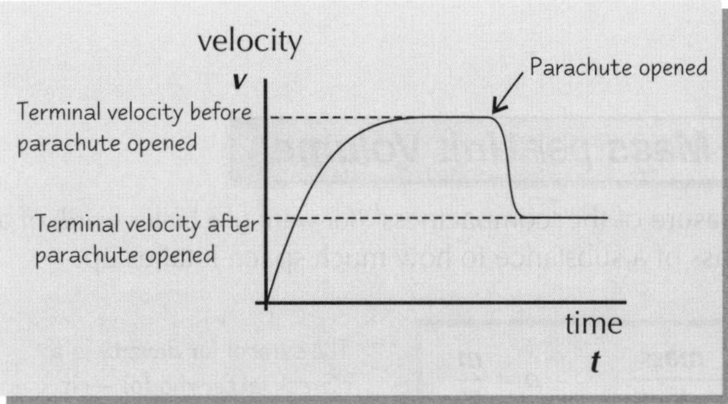

Practice Questions

Q1 What forces limit the speed of a skier going down a slope?

Q2 What conditions cause a terminal velocity to be reached?

Q3 Sketch a graph to show how the velocity changes with time for an object falling through air.

Exam Question

Q1 A space probe free-falls towards the surface of a planet.
The graph on the right shows the velocity data recorded by the probe as it falls.

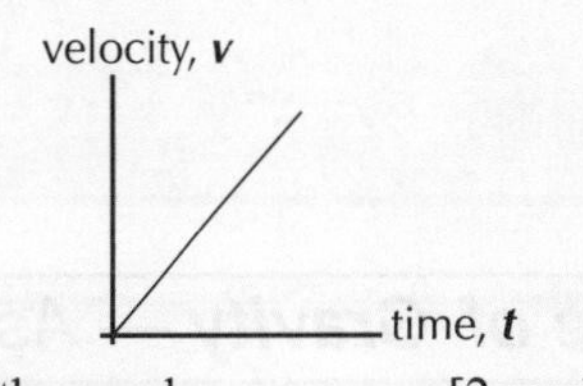

(a) The planet does not have an atmosphere. Explain how you can tell this from the graph. [2 marks]

(b) Sketch the velocity-time graph you would expect to see if the planet did have an atmosphere. [2 marks]

(c) Explain the shape of the graph you have drawn. [3 marks]

You'll never understand this without going parachuting...*

When you're doing questions about terminal velocity, remember the frictional forces reduce acceleration, not speed. They usually don't slow an object down, apart from in the parachute example, where the skydiver is travelling faster when the parachute opens than the terminal velocity for the parachute-skydiver combination.

* No. 37 in a series of the 100 least convincing excuses for an interesting holiday.

Mass, Weight and Centre of Gravity

I'm sure you know all this 'mass', 'weight' and 'density' stuff from GCSE. But let's just make sure...

The Mass of a Body makes it Resist Changes in Motion

1) The **mass** of an object is the **amount of 'stuff'** (or **matter**) in it. It's measured in **kg**.
2) The greater an object's mass, the greater its **resistance** to a **change in velocity** (called its **inertia**).
3) The **mass** of an object **doesn't change** if the strength of the **gravitational field** changes.
4) Weight is a **force**. It's measured in **newtons** (N), like all forces.
5) Weight is the **force experienced by a mass** due to a **gravitational field.**
6) The weight of an object **does vary** according to the size of the **gravitational field** acting on it.

weight = mass × gravitational field strength (W = mg) where $g = 9.81\ \text{Nkg}^{-1}$ on Earth.

This table shows Gerald (the lion*)'s mass and weight on the Earth and the Moon.

Name	Quantity	Earth ($g = 9.81\ \text{Nkg}^{-1}$)	Moon ($g = 1.6\ \text{Nkg}^{-1}$)
Mass	Mass (scalar)	150 kg	150 kg
Weight	Force (vector)	1471.5 N	240 N

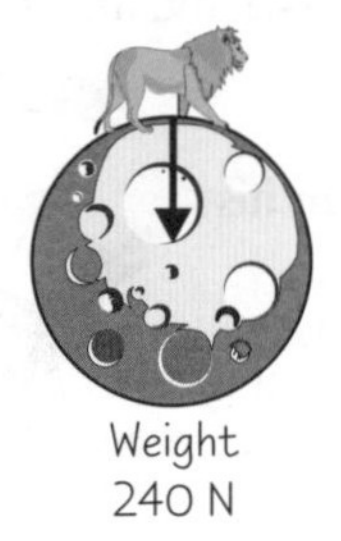

Density is Mass per Unit Volume

Density is a measure of the 'compactness' (for want of a better word) of a substance.
It relates the mass of a substance to how much space it takes up.

$$\textit{density} = \frac{\textit{mass}}{\textit{volume}} \qquad \rho = \frac{m}{V}$$

The symbol for density is a Greek letter rho (ρ) — it looks like a p but it isn't.

The **units** of **density** are $\mathbf{g\,cm^{-3}}$ or $\mathbf{kg\,m^{-3}}$
N.B. $1\ \text{g cm}^{-3} = 1000\ \text{kg m}^{-3}$

1) The density of an object depends on what it's made of. Density **doesn't vary** with **size or shape**.
2) The **average density** of an object determines whether it **floats** or **sinks**.
3) A solid object will **float** on a fluid if it has a **lower density** than the **fluid**.

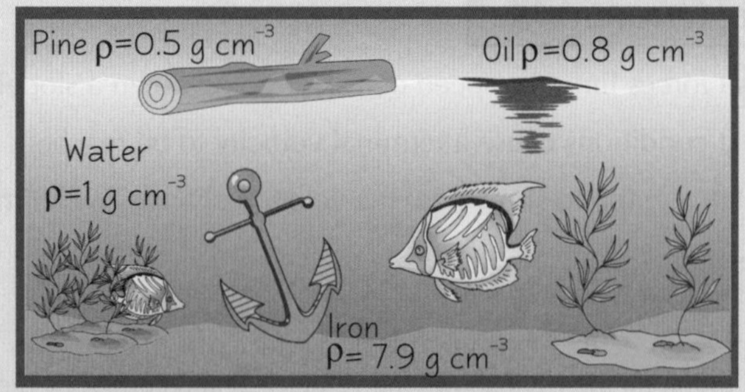

Centre of Gravity — Assume All the Mass is in One Place

1) The **centre of gravity** (or centre of mass) of an object is the **single point** that you can consider its **whole weight** to **act through** (whatever its orientation).
2) The object will always **balance** around this **point**, although in some cases the **centre of gravity** will **fall outside** the object.

*Yes, I know — I just like lions, OK...

Mass, Weight and Centre of Gravity

Find the *Centre of Gravity* either by *Symmetry* or *Experiment*

Experiment to find the Centre of Gravity of an Irregular Object

1) Hang the object freely from a point (e.g. one corner).
2) Draw a vertical line downwards from the point of suspension — use a plumb bob to get your line exactly vertical.
3) Hang the object from a different point.
4) Draw another vertical line down.
5) The centre of gravity is where the two lines cross.

string
drawn vertical line
clip
drawn vertical line
Centre of gravity

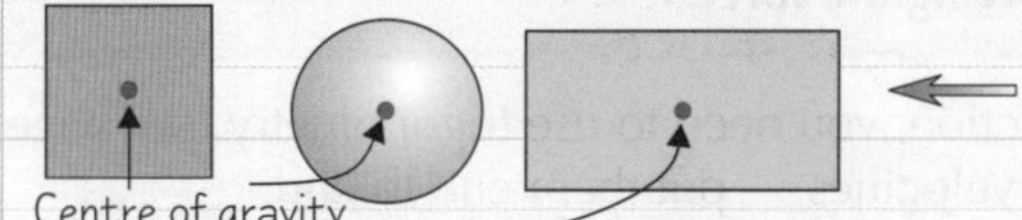

For a regular object you can just use symmetry. The centre of gravity of any regular shape is at its centre.

How High the *Centre of Gravity* is tells you *How Stable* the Object is

1) An object will be nice and **stable** if it has a **low centre** of **gravity** and a **wide base area**. This idea is used a lot in design, e.g. Formula 1 racing cars.

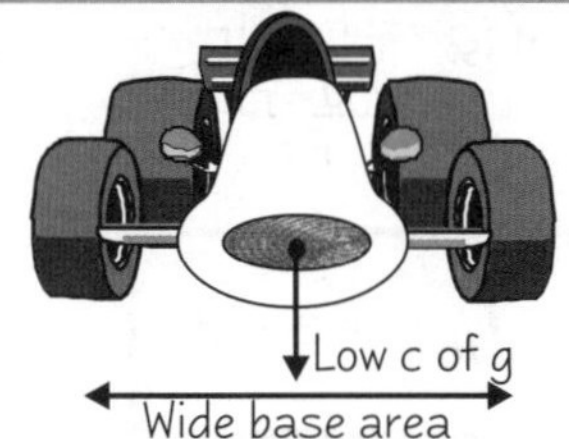

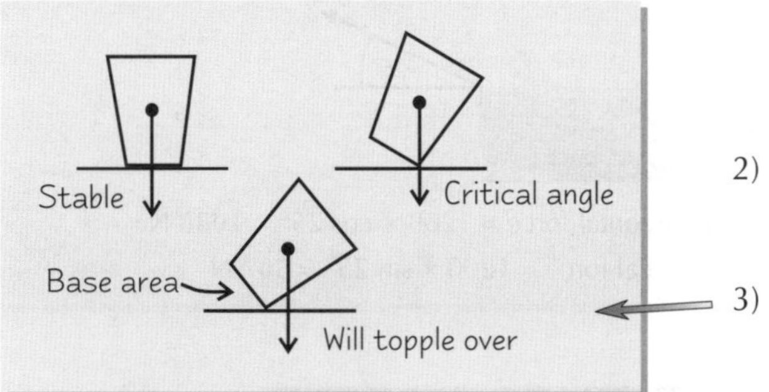

2) The **higher** the **centre of gravity**, and the **smaller** the **base area**, the **less stable** the object will be. Think of unicyclists...
3) An object will topple over if a **vertical line** drawn **downwards** from its **centre of gravity** falls **outside** its **base area**.

Practice Questions

Q1 A lioness has a mass of 200 kg. What would be her mass and weight on the Earth (where $g = 9.8\ \text{Nkg}^{-1}$) and on the Moon (where $g = 1.6\ \text{Nkg}^{-1}$)?

Q2 What is meant by the centre of gravity of an object?

Exam Questions

Q1 (a) Define **density**. [1 mark]

(b) A cylinder of aluminium, radius 4 cm and height 6 cm, has a mass of 820 g. Calculate its density. [3 marks]

(c) Use the information from part (b) to calculate the mass of a cube of aluminium of side 5 cm. [1 mark]

Q2 Describe an experiment to find the centre of gravity of an object of uniform density with a constant thickness and irregular cross-section. Identify one major source of uncertainty and suggest a way to reduce its effect on the accuracy of your result. [5 marks]

The centre of gravity of this book should be round about page 47...

This is a really useful area of physics. To would-be nuclear physicists it might seem a little dull, but if you want to be an engineer — something a bit more useful (no offence Einstein) — then things like centre of gravity and density are dead important things to understand. You know, for designing things like cars and submarines... yep, pretty useful I'd say.

Forces and Equilibrium

Remember the vector stuff from the beginning of the section... good, you're going to need it...

Resolving a Force means Splitting it into Components

1) **Forces** are **vector quantities** and so when you draw the forces on an object, the **arrow labels** should show the **size** and **direction** of the forces.
2) Forces can be in **any direction**, so they're not always at right angles to each other. This is sometimes a bit **awkward** for **calculations**.
3) To make an 'awkward' force easier to deal with, you can think of it as **two separate forces**, acting at **right angles** to **each other**.

The force F has exactly the same effect as the horizontal and vertical forces, F_H and F_V.
Replacing F with F_H and F_V is called **resolving the force** F.

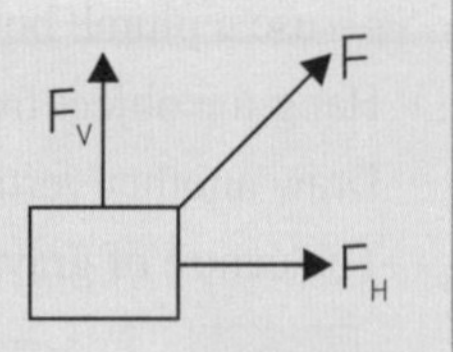

4) To find the size of a component force in a particular direction, you need to use trigonometry (see page 5). Forces are vectors, so you treat them in the same way as velocities — put them end to end.

So this...

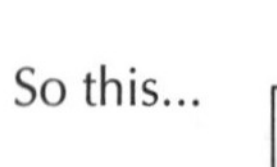

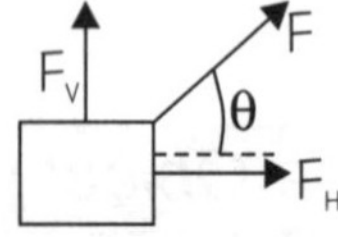

...could be drawn like this:

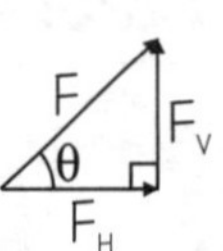

Using trig. you get:

$$\frac{F_H}{F} = \cos\theta \quad \text{or} \quad F_H = F\cos\theta$$

And:

$$\frac{F_V}{F} = \sin\theta \quad \text{or} \quad F_V = F\sin\theta$$

Example

A tree trunk is pulled along the ground by an elephant exerting a force of 1200 N at an angle of 25° to the horizontal. Calculate the components of this force in the horizontal and vertical directions.

Horizontal force = 1200 × cos 25° = **1088 N**
Vertical force = 1200 × sin 25° = **507 N**

Three Forces Acting on a Point in Equilibrium form a Triangle

1) When **three** forces all act on an object in **equilibrium**, you know there is **no net force** on the object — the sum of the forces is zero.

$$F_1 + F_2 + F_3 = 0$$

2) You can draw the forces as a triangle, forming a **closed loop** like these:
3) Be careful when you draw the triangles not to go into autopilot and draw F_3 as the sum of F_1 and F_2 — it has to be in the **opposite** direction to balance the other two forces.

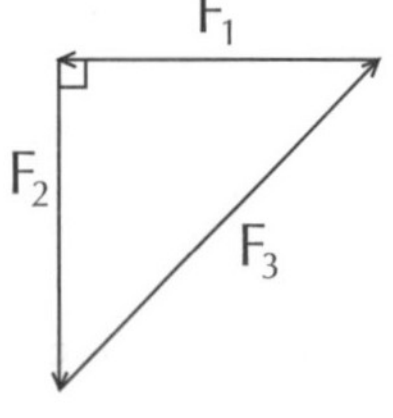

4) You can then use the triangles to while away Saturday nights in with your friends working out the magnitude or direction of a missing force. What could be more fun?

Example

Tim hangs a picture of his brother up using a piece of string. The picture weighs 0.3 N and is in equilibrium, as shown. Find the magnitude of ***F***.

Draw a vector triangle.

Then you need to use **Pythagoras** to find the magnitude of F.

$$F = \sqrt{0.3^2 + 0.4^2} = \sqrt{0.25^2} = \mathbf{0.5\ N}$$

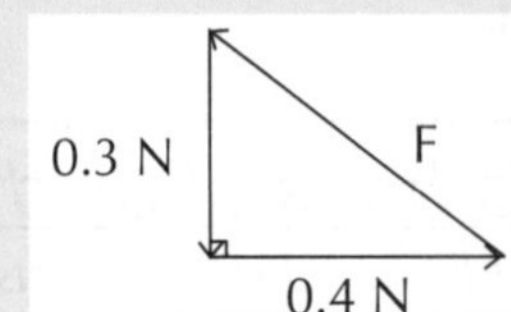

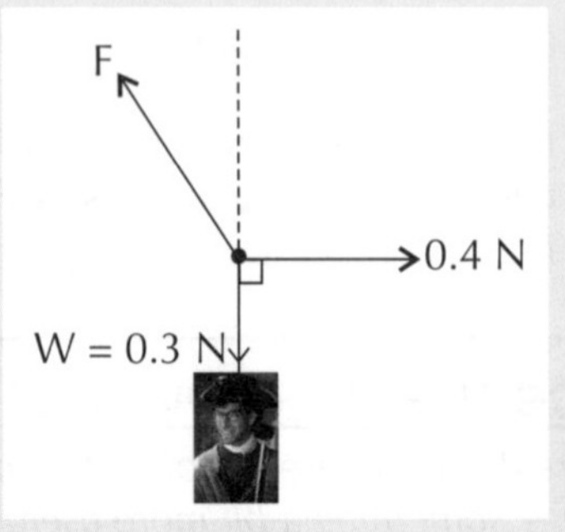

Forces and Equilibrium

You Add the Components Back Together to get the Resultant Force

1) If **two forces** act on an object, you find the **resultant** (total) **force** by adding the **vectors** together and creating a **closed triangle**, with the resultant force represented by the **third side**.
2) Forces are vectors (as you know), so you use **vector addition** — draw the forces as vector arrows put 'tail to top'.
3) Then it's yet more trigonometry to find the **angle** and the **length** of the third side.

Example

Two dung beetles roll a dung ball along the ground at constant velocity. Beetle A applies a force of 0.5 N northwards while beetle B exerts a force of only 0.2 N eastwards. What is the resultant force on the dung ball?

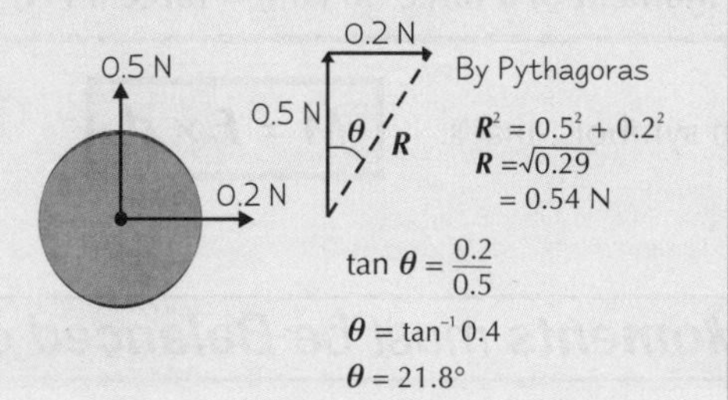

The resultant force is **0.54 N** at an angle of **21.8°** from North.

Choose sensible Axes for Resolving

Use directions that **make sense** for the situation you're dealing with. If you've got an object on a slope, choose your directions **along the slope** and **at right angles to it**. You can turn the paper to an angle if that helps.

Always choose sensible axes

Examiners like to call a slope an "inclined plane".

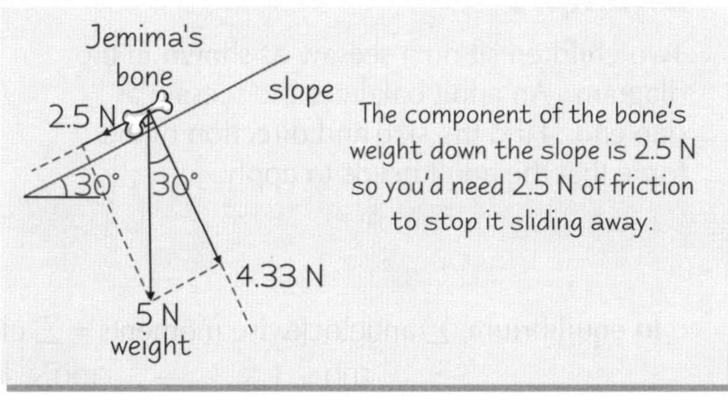

The component of the bone's weight down the slope is 2.5 N so you'd need 2.5 N of friction to stop it sliding away.

Practice Questions

Q1 Sketch a free-body force diagram for an ice hockey puck moving across the ice (assuming no friction).

Q2 What are the horizontal and vertical components of the force F?

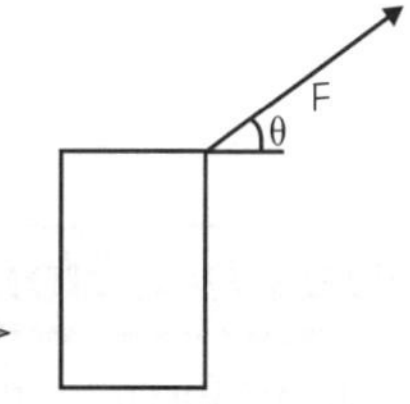

Exam Questions

Q1 A picture is suspended from a hook as shown in the diagram. Calculate the tension force, *T*, in the string.

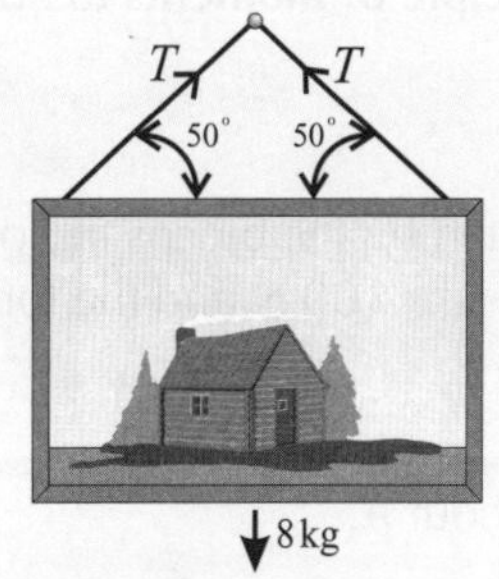

[2 marks]

Q2 Two elephants pull a tree trunk as shown in the diagram. Calculate the resultant force on the tree trunk.

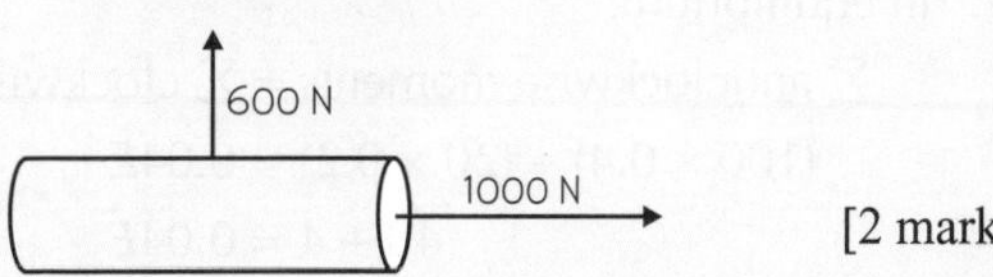

[2 marks]

Free-body force diagram — sounds like something you'd get with a dance mat...

*Remember those F cos θ and F sin θ bits. Write them on bits of paper and stick them to your wall. Scrawl them on your pillow. Tattoo them on your brain. Whatever it takes — you just **have to learn them**.*

Moments and Torques

*This is not a time for jokes. There is not a moment to lose. The time for torquing is over. Oh ho ho ho ho *bang*. (Ow.)*

A Moment is the Turning Effect of a Force

The **moment**, or **torque**, of a **force** depends on the **size** of the force and **how far** the force is applied from the **turning point**:

moment of a force (in Nm) = **force** (in N) × **perpendicular distance from pivot** (in m)

In symbols, that's: $M = F \times d$

Moments must be Balanced or the Object will Turn

The **principle of moments** states that for a body to be in **equilibrium**, the **sum of the clockwise moments** about any point **equals** the **sum of the anticlockwise moments** about the same point.

Example

Two children sit on a seesaw as shown in the diagram. An adult balances the seesaw at one end. Find the size and direction of the force that the adult needs to apply.

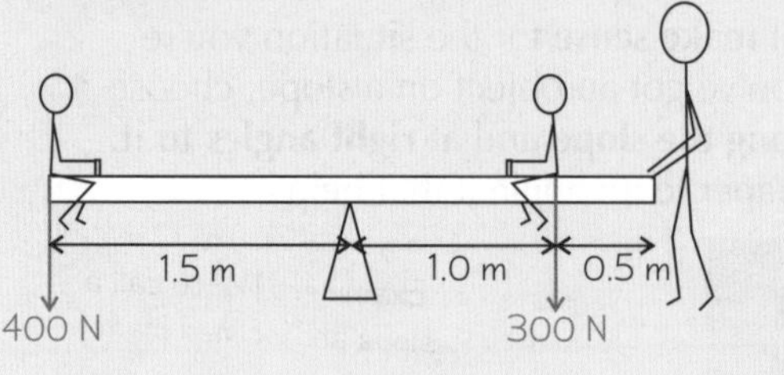

In equilibrium, Σ anticlockwise moments = Σ clockwise moments

$400 \times 1.5 = 300 \times 1 + 1.5F$

$600 = 300 + 1.5F$

Final answer: F = 200 N downwards

Σ means "the sum of"

Muscles, Bones and Joints Act as Levers

1) In a lever, an **effort force** (in this case from a muscle) acts against a **load force** (e.g. the weight of your arm) by means of a **rigid object** (the bone) rotating around a **pivot** (the joint).
2) You can use the **principle of moments** to answer lever questions:

Example

Find the force exerted by the biceps in holding a bag of gold still. The bag of gold weighs 100 N and the forearm weighs 20 N.

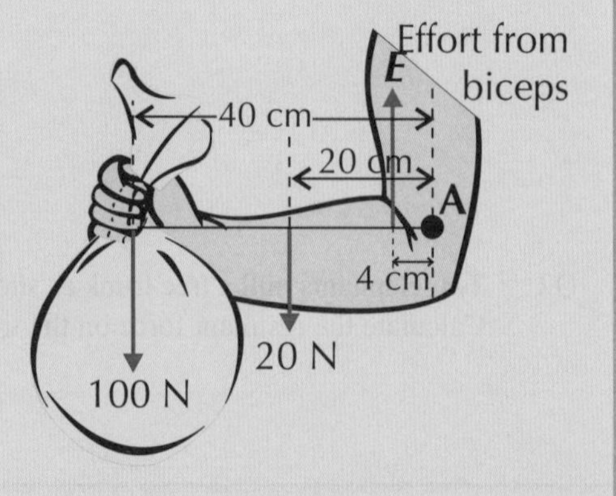

Take moments about **A**.

In equilibrium:

Σ anticlockwise moments = Σ clockwise moments

$(100 \times 0.4) + (20 \times 0.2) = 0.04E$

$40 + 4 = 0.04E$

Final answer: E = 1100 N

Moments and Torques

A Couple is a Pair of Forces

1) A couple is a **pair** of **forces** of **equal size** which act **parallel** to each other, but in **opposite directions**.
2) A couple doesn't cause any resultant linear force, but **does** produce a **turning force** (usually called a **torque** rather than a moment).

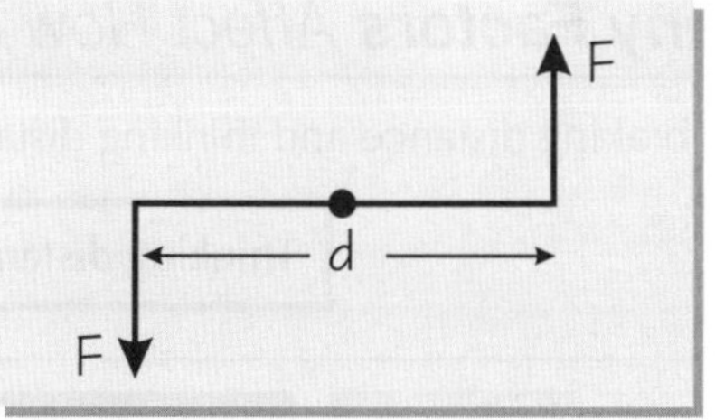

The **size** of this **torque** depends on the **size** of the **forces** and the **distance** between them.

Torque of a couple (in Nm) = **size of one of the forces** (in N) × **perpendicular distance between the forces** (in m)

In symbols, that's: $T = F \times d$

Example

A cyclist turns a sharp right corner by applying equal but opposite forces of 20 N to the ends of the handlebars.

The length of the handlebars is 0.6 m.
Calculate the torque applied to the handlebars.

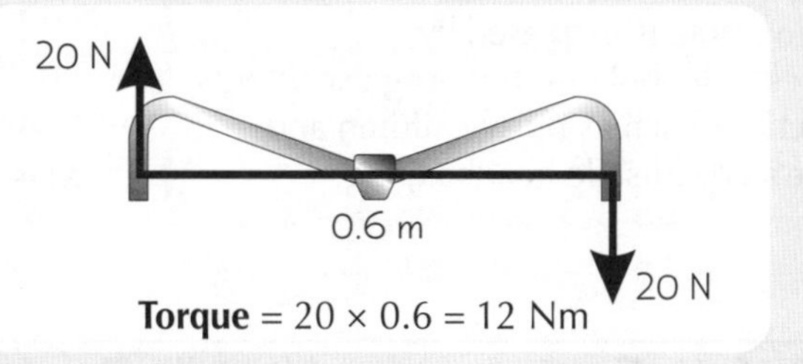

Torque = 20 × 0.6 = 12 Nm

Practice Questions

Q1 A girl of mass 40 kg sits 1.5 m from the middle of a seesaw.
Show that her brother, mass 50 kg, must sit 1.2 m from the middle if the seesaw is to balance.

Q2 What is meant by the word 'couple'?

Exam Questions

Q1 A driver is changing his flat tyre. The torque required to undo the nut is 60 Nm.
He uses a 0.4 m long double-ended wheel wrench.
Calculate the force that he must apply at each end of the spanner. [2 marks]

Q2 A diver of mass 60 kg stands on the end of a diving board 2 m from the pivot point.
Calculate the upward force exerted on the retaining spring 30 cm from the pivot.

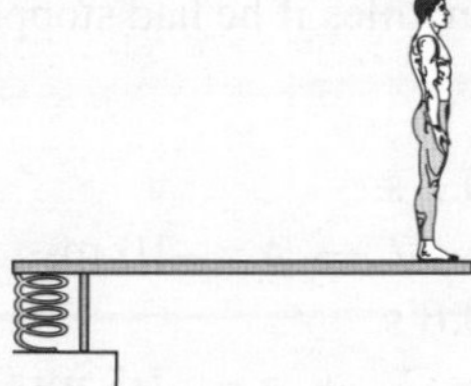

[2 marks]

It's all about balancing — just ask a tightrope walker...

*They're always seesaw questions aren't they. It'd be nice if once, **just once**, they'd have a question on... I don't know, rotating knives or something. Just something unexpected... anything. It'd make physics a lot more fun, I'm sure. *sigh**

Car Safety

Some real applications now — how to avoid collisions, and how car manufacturers try to make sure you survive.

Many Factors Affect How *Quickly* a Car *Stops*

The braking distance and thinking distance together make the **total distance you need to stop** after you see a problem:

Thinking distance + Braking distance = Stopping distance

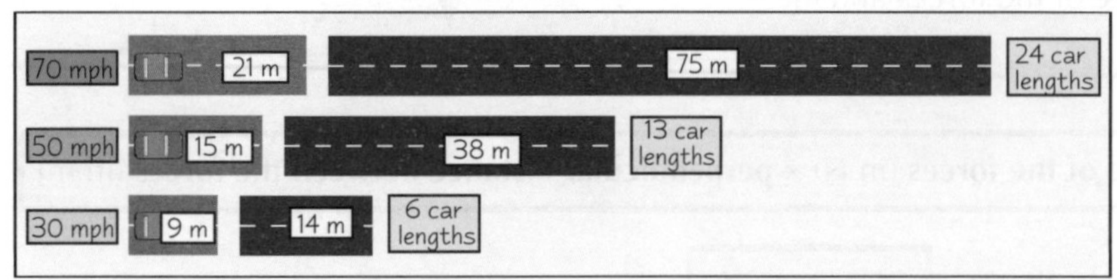

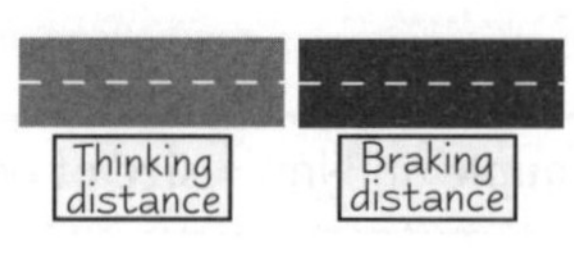

In an exam you might need to list factors that affect the thinking and braking distances.

thinking distance = speed × reaction time

Reaction time is increased by **tiredness**, **alcohol** or other **drug** use, **illness**, **distractions** such as noisy children and Wayne's World-style headbanging.

Braking distance depends on the **braking force, friction** between the tyres and the road, the **mass** and the **speed**.

Braking force is reduced by **reduced friction** between the brakes and the wheels (**worn** or **badly adjusted brakes**).

Friction between the tyres and the road is reduced by **wet** or **icy** roads, **leaves or dirt** on the road, **worn-out tyre treads**, etc.

Mass is affected by the size of the car and what you put in it.

Car Safety Features are Usually *Designed* to *Slow You Down Gradually*

Modern cars have **safety features** built in. Many of them make use of the idea of slowing the collision down so it **takes you longer to stop**, so your **deceleration is less** and there is **less force** on you.

Safety features you need to know about are:

1) **Seatbelts** keep you in your seat and also 'give' a little so that you're brought to a stop over a longer time.
2) **Airbags** inflate when you have a collision and are big and squishy so they stop you hitting hard things and slow you down gradually. (More about airbags and how they work on the next page.)
3) **Crumple zones** at the front and back of the car are designed to give way more easily and absorb some of the energy of the collision.
4) **Safety cages** are designed to prevent the area around the occupants of the car from being crushed in.

Example

Giles's car bumps into the back of a stationary bus. The car was travelling at 2 ms^{-1} and comes to a stop in 0.2 s. Giles was wearing his seatbelt and takes 0.8 s to stop. The mass of the car is 1000 kg and Giles's mass is 75 kg.

a) Find the decelerations of Giles and the car.
b) Calculate the average force acting on Giles during the accident.
c) Work out the average force that would have acted on Giles if he had stopped in as short a time as the car.

a) Use $\boldsymbol{v} = \boldsymbol{u} + \boldsymbol{at}$:
For the car: $\boldsymbol{u} = 2\ ms^{-1}$, $\boldsymbol{v} = 0$, $\boldsymbol{t} = 0.2$ s
Which gives: $0 = 2 + 0.2\boldsymbol{a} \Rightarrow 0.2\boldsymbol{a} = -2 \Rightarrow \boldsymbol{a} = -10\ ms^{-2}$ so the **deceleration = 10 ms^{-2}**
For Giles: $\boldsymbol{u} = 2\ ms^{-1}$, $\boldsymbol{v} = 0$, $\boldsymbol{t} = 0.8$ s
Which gives: $0 = 2 + 0.8\boldsymbol{a} \Rightarrow 0.8\boldsymbol{a} = -2 \Rightarrow \boldsymbol{a} = -2.5\ ms^{-2}$ so the **deceleration = 2.5 ms^{-2}**

b) Use $\boldsymbol{F} = \boldsymbol{ma} = 75 \times 2.5 =$ **187.5 N**

c) Use $\boldsymbol{F} = \boldsymbol{ma}$ again, but with 10 ms^{-2} instead of 2.5 ms^{-2}: $\boldsymbol{F} = \boldsymbol{ma} = 75 \times 10 =$ **750 N**

Car Safety

Airbags are Triggered by Rapid Deceleration

It's pretty hard going trying to get an airbag to go off when you need it — and not end up having a face full of airbag every time you stop at traffic lights. Here's how they work...

1) All airbags are **triggered to inflate using** sensors that detect the **rapid deceleration** of a car in a crash.
2) Most cars use a microchip **accelerometer** — where **rapid deceleration** changes the **capacitance** of part of the microchip. This change can be detected by the microchip's electronics, which send an "inflate now" signal to the airbag modules in the car.
3) This kicks off a rapid **chemical reaction** that produces a load of inert gas to inflate the air bag.
4) It's a lot to do in a very **short space of time** if it's going to get to your head before you get up close and personal with a steering wheel. Airbags inflate in less than a tenth of a second once triggered.
5) As soon as they're inflated, the airbags begin to deflate as gas escapes through flaps in the fabric.

GPS Devices Find Where You Are Using Trilateration

Some cars are fitted with glitzy global positioning systems (**GPS**) to help you find out **where** you are. They do this using a process called **trilateration**...

1) The GPS in your car receives signals from at least **three satellites**, each transmitting their **location** and the **time** the signal was sent.
2) As the signals take a short amount of time to reach the GPS, there is a short **delay** between the time sent and the time each signal is received. The further away the satellite, the longer it takes the signal to get to your car.
3) By knowing the **time delay** for each satellite signal, the GPS can calculate the **distance** to each satellite. You then know that you must be somewhere on the surface of a sphere that's centred on that satellite.
4) If you know the distances to **three** satellites, you must be at the point where all three spheres **meet**. Clever, huh?
5) GPS systems actually use at least **four** satellites to locate you. You only need three, but the more satellites the more **accurately** you can know your position.

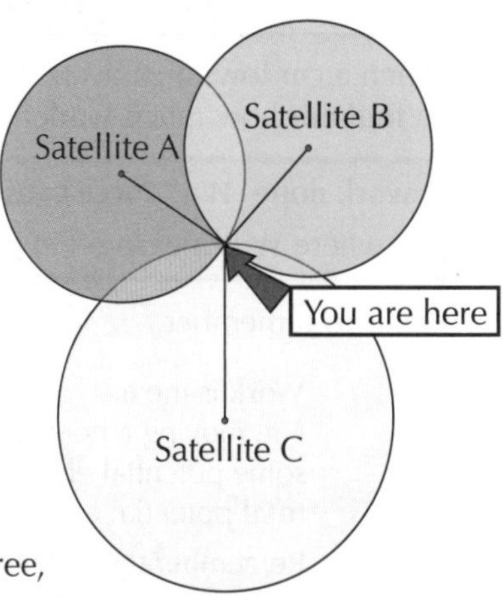

Practice Questions

Q1 What equation can you use to work out the force you experience during a collision?

Q2 What factor affects both thinking distance and braking distance?

Q3 Describe how airbags are triggered to inflate when a car is in a collision.

Exam Questions

Q1 Sarah sees a cow step into the road 30 m ahead of her. Sarah's reaction time is 0.5 s. She is travelling at 20 ms^{-1}. Her maximum braking force is 10 000 N and her car (with her in it) has a mass of 850 kg.

(a) How far does she travel before applying her brakes? [2 marks]

(b) Calculate Sarah's braking distance. Assume she applies the maximum braking force until she stops. [3 marks]

(c) Does Sarah hit the cow? Justify your answer with a suitable calculation. [1 mark]

Q2 In a crash test a car slams into a solid barrier at 20 ms^{-1}. The car comes to a halt in 0.1 s. The crash test dummy goes through the windscreen and hits the barrier at a speed of 18 ms^{-1} and then also comes to a stop in 0.1 s. The mass of the car is 900 kg and the mass of the dummy is 50 kg.

(a) Calculate the forces on the car and the dummy as they are brought to a stop. [4 marks]

(b) The car is modified to include crumple zones and an airbag. Explain what difference this will make and why. [3 marks]

Crumple zone — the heap of clothes on my bedroom floor...

Being safe in a car is mainly common sense — don't drive if you're ill, drunk or just tired and don't drive a car with dodgy brakes. But you still need to cope with exam questions, so don't go on till you're sure you know this all off by heart.

Work and Power

As everyone knows, work in Physics isn't like normal work. It's harder. Work also has a specific meaning that's to do with movement and forces. You'll have seen this at GCSE — it just comes up in more detail for AS level.

Work is Done Whenever Energy is Transferred

This table gives you some examples of **work being done** and the **energy changes** that happen.

1) Usually you need a force to move something because you're having to **overcome another force**.
2) The thing being moved has **kinetic energy** while it's **moving**.
3) The kinetic energy is transferred to **another form of energy** when the movement stops.

ACTIVITY	WORK DONE AGAINST	FINAL ENERGY FORM
Lifting up a box.	gravity	gravitational potential energy
Pushing a chair across a level floor.	friction	heat
Pushing two magnetic north poles together.	magnetic force	magnetic energy
Stretching a spring.	stiffness of spring	elastic potential energy

The word **'work'** in Physics means the **amount of energy transferred** from one form to another when a force causes a movement of some sort.

Work = Force × Distance

When a car tows a caravan, it applies a force to the caravan to moves it to where it's wanted.
To **find out** how much **work** has been **done**, you need to use the **equation**:

work done (W) = force causing motion (F) × distance moved (s)
...where W is measured in joules (J), F is measured in newtons (N) and s is measured in metres (m).

Points to remember:

1) **Work** is the **energy** that's been **changed** from one form to another — it's not necessarily the **total** energy. E.g. moving a book from a low shelf to a higher one will increase its gravitational potential energy, but it had some potential energy to start with. Here, the **work done** would be the **increase** in potential energy, **not the total** potential energy.
2) Remember the distance needs to be measured in metres — if you have **distance in centimetres or kilometres**, you need to **convert** it to metres first.
3) The force F will be a **fixed** value in any calculations, either because it's **constant** or because it's the **average** force.
4) The equation assumes that the **direction of the force** is the **same** as the **direction of movement**.
5) The equation gives you the **definition** of the joule (symbol J): 'One joule is the work done when a force of 1 newton moves an object through a distance of 1 metre'.

The Force isn't always in the Same Direction as the Movement

Sometimes the **direction of movement** is **different** from the **direction of the force**.

Example

1) To **calculate the work done** in a situation like the one on the right, you need to consider the **horizontal** and **vertical components** of the **force**.
2) The only **movement** is in the **horizontal** direction. This means the **vertical force** is not causing any motion (and hence not doing any work) — it's just **balancing** out some of the **weight**, meaning there's a **smaller reaction force**.

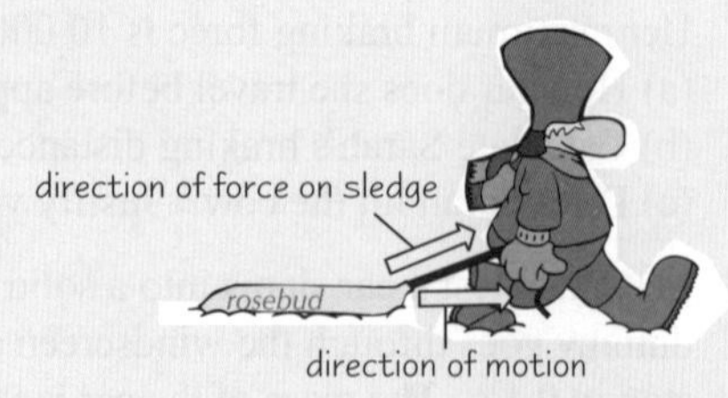

3) The horizontal force is causing the motion — so to **calculate** the **work done**, this is the **only force** you need to consider. Which means we get:

$$W = Fs\cos\theta$$

Where θ is the **angle** between the **direction of the force** and the **direction of motion**. See page 22 for more on resolving forces.

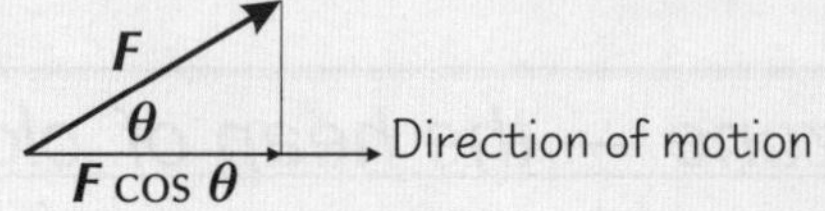

Work and Power

Power = Work Done per Second

Power means many things in everyday speech, but in physics (of course!) it has a special meaning. Power is the **rate of doing work** — in other words it is the **amount of energy transformed** from one form to another **per second.** You **calculate power** from this equation:

> **Power (*P*) = work done (*W*) / time (*t*)**
> ...where ***P*** is measured in watts (W), ***W*** is measured in joules (J) and ***t*** is measured in seconds (s)

The **watt** (symbol W) is defined as a **rate of energy transfer** equal to **1 joule per second** (Js^{-1}).
Yep, that's another **equation and definition** for you to **learn**.

Power is also Force × Velocity (P = Fv)

Sometimes, it's **easier** to use **this version** of the power equation. This is how you get it:

1) You **know** $P = W/t$.
2) You also **know** $W = Fs$, which gives $P = Fs/t$.
3) But $v = s/t$, which you can substitute into the above equation to give $P = Fv$.
4) It's easier to use this if you're given the **speed** in the question.
   Learn this equation as a **shortcut** to link **power** and **speed**.

Example

A car is travelling at a speed of $10\ ms^{-1}$ and is kept going against the frictional force by a driving force of 500 N in the direction of motion. Find the power supplied by the engine to keep the car moving.

Use the shortcut $P = Fv$, which gives:
$P = 500 \times 10 = 5000$ W

If the force and motion are in different directions, you can replace F with $F\cos\theta$ to get: $P = Fv\cos\theta$

You **aren't** expected to **remember** this equation, but it's made up of bits that you **are supposed to know**, so be ready for the possibility of calculating **power** in a situation where the **direction of the force and direction of motion are different**.

Practice Questions

Q1 Write down the equation used to calculate work if the force and motion are in the same direction.

Q2 Write down the equation for work if the force is at an angle to the direction of motion.

Q3 Write down the equations relating (i) power and work and (ii) power and speed.

Exam Questions

Q1 A traditional narrowboat is drawn by a horse walking along the towpath.
The horse pulls the boat at a constant speed between two locks which are 1500 m apart. The tension in the rope is 100 N at 40° to the direction of motion.

(a) How much work is done on the boat? [2 marks]
(b) The boat moves at $0.8\ ms^{-1}$. Calculate the power supplied to the boat in the direction of motion. [2 marks]

Q2 A motor is used to lift a 20 kg load a height of 3 m. (Take $g = 9.81\ Nkg^{-1}$.)
(a) Calculate the work done in lifting the load. [2 marks]
(b) The speed of the load during the lift is $0.25\ ms^{-1}$. Calculate the power delivered by the motor. [2 marks]

Work — there's just no getting away from it...

Loads of equations to learn. Well, that's what you came here for, after all. Can't beat a good bit of equation-learning, as I've heard you say quietly to yourself when you think no one's listening. Aha, can't fool me. Ahahahahahahahahahahahahaha.

Energy *can never be* ***lost****. I repeat — **energy** can **never** be lost. Which is basically what I'm about to take up two whole pages saying. But that's, of course, because you need to do exam questions on this as well as understand the principle.*

Learn the **Principle** of **Conservation** of **Energy**

The **principle of conservation of energy** says that:

> Energy **cannot be created** or **destroyed**. Energy **can be transferred** from one form to another but the total amount of energy in a closed system will not change.

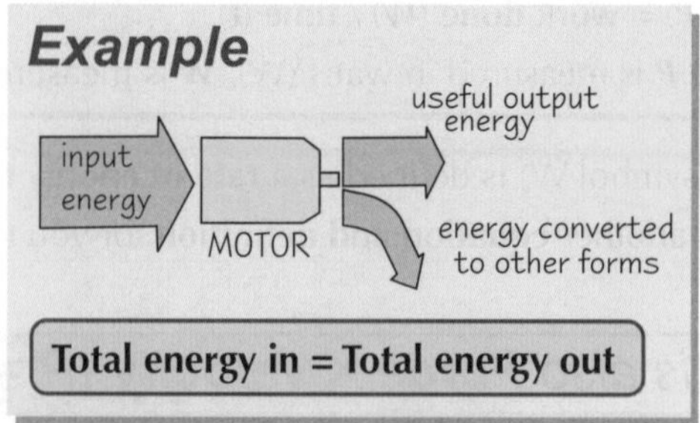

You need it for **Questions** about **Kinetic** and **Potential Energy**

The principle of conservation of energy nearly always comes up when you're doing questions about changes between kinetic and potential energy.

A quick reminder:

1) **Kinetic energy** is energy of anything **moving**, which you work out from $E_k = \frac{1}{2}mv^2$, where v is the velocity it's travelling at and m is its mass.
2) There are **different types of potential energy** — e.g. gravitational and elastic.
3) **Gravitational potential energy** is the energy something gains if you lift it up. You work it out using: $\Delta E_p = mg\Delta h$, where m is the mass of the object, Δh is the height it is lifted and g is the gravitational field strength (9.81 Nkg^{-1} on Earth).
4) **Elastic potential energy** (elastic stored energy) is the energy you get in, say, a stretched rubber band or spring. You work this out using $E = \frac{1}{2}ke^2$, where **e** is the extension of the spring and k is the stiffness constant.

Examples These pictures show you three **examples** of changes between kinetic and potential energy.

1) As Becky throws the **ball upwards, kinetic energy** is converted into **gravitational potential energy**. When it **comes down** again, that **gravitational potential** energy is **converted back** into **kinetic** energy.
2) As Dominic goes **down the slide, gravitational potential energy** is converted to **kinetic energy**.
3) As Simon bounces upwards from the trampoline, **elastic potential energy** is converted to **kinetic energy**, to **gravitational potential energy**. As he comes back down again, that **gravitational potential** energy is **converted back** to **kinetic** energy, to **elastic potential** energy, and so on.

In **real life** there are also **frictional forces** — Simon would have to use some **force** from his **muscles** to keep **jumping** to the **same height** above the trampoline each time. Each time the trampoline **stretches**, some **heat** is generated in the trampoline material. You're usually told to **ignore friction** in exam questions — this means you can **assume** that the **only forces** are those that provide the **potential or kinetic energy** (in this example that's **Simon's weight** and the **tension** in the springs and trampoline material).
If you're ignoring friction, you can say that the **sum of the kinetic and potential energies is constant.**

Conservation of Energy

Use Conservation of Energy to Solve Problems

You need to be able to **use** conservation of mechanical energy (change in potential energy = change in kinetic energy) to solve problems. The classic example is the **simple pendulum**.

In a simple pendulum, you assume that all the mass is in the **bob** at the end.

Example

A simple pendulum has a mass of 700 g and a length of 50 cm. It is pulled out to an angle of 30° from the vertical.

(a) Find the gravitational potential energy stored in the pendulum bob.

Start by drawing a diagram.

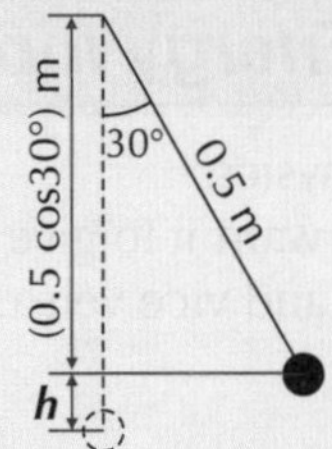

You can work out the increase in height, ***h***, of the end of the pendulum using trig.

Gravitational potential energy = mgh

$= 0.7 \times 9.81 \times (0.5 - 0.5\cos 30°)$

$= 0.46$ J

(b) The pendulum is released. Find the maximum speed of the pendulum bob as it passes the vertical position.

To find the *maximum* speed, assume no air resistance, then $mgh = \frac{1}{2}mv^2$.

So $\frac{1}{2}mv^2 = 0.46$

rearrange to find $v = \sqrt{\frac{2 \times 0.46}{0.7}} = 1.15 \text{ ms}^{-1}$

OR

Cancel the ms and rearrange to give:

$v^2 = 2gh$

$= 2 \times 9.81 \times (0.5 - 0.5\cos 30°)$

$= 1.31429...$

$v = 1.15 \text{ ms}^{-1}$

You could be asked to apply this stuff to just about any situation in the exam. **Rollercoasters** are a bit of a favourite.

Practice Questions

Q1 State the principle of conservation of energy.

Q2 What are the equations for calculating kinetic energy and gravitational potential energy?

Q3 Show that, if there's no air resistance and the mass of the string is negligible, the speed of a pendulum is independent of the mass of the bob.

Exam Questions

Q1 A skateboarder is on a half-pipe. He lets the board run down one side of the ramp and up the other. The height of the ramp is 2 m. Take **g** as 9.81 Nkg^{-1}.

(a) If you assume that there is no friction, what would be his speed at the lowest point of the ramp? [3 marks]

(b) How high will he rise up the other side? [1 mark]

(c) Real ramps are not frictionless, so what must the skater do to reach the top on the other side? [1 mark]

Q2 A 20 g rubber ball is released from a height of 8 m. (Assume that the effect of air resistance is negligible.)

(a) Find the kinetic energy of the ball just before it hits the ground. [2 marks]

(b) The ball strikes the ground and rebounds to a height of 6.5 m. How much energy is converted to heat and sound in the impact with the ground? [2 marks]

Energy is never lost — it just sometimes prefers the scenic route...

Remember to check your answers — I can't count the number of times I've forgotten to square the velocities or to multiply by the ½... I reckon it's definitely worth the extra minute to check.

Energy, it seems, is pretty reliable stuff — whatever you do, you can't create or destroy it — it'll always be there. But, like homework and small children, it's very easy to lose — you know it's there somewhere, you just can't find it.

All Energy Transfers Involve Losses

You saw on the last page that **energy can never be created or destroyed**. But whenever **energy** is **converted** from one form to another, some is always **'lost'**. It's still there (i.e. it's **not destroyed**) — it's just not in a form you can **use**.

Most often, **energy** is lost as **heat** — e.g. **computers** and **TVs** are always **warm** when they've been on for a while. In fact, **no device** (except possibly a heater) is ever **100% efficient** (see below) because some energy is **always** lost as **heat**. (You want heaters to give out heat, but in other devices the heat loss isn't useful.) Energy can be **lost** in other forms too (e.g. **sound**) — the important thing is the lost energy **isn't** in a **useful** form and you **can't** get it back.

Efficiency is the Ratio of Useful Energy Output to Total Energy Input

Efficiency is one of those words we use all the time, but it has a **specific meaning** in Physics.
It's a measure of how well a **device** converts the **energy** you put **in** into the energy you **want** it to give **out**.
So, a device that **wastes** loads of **energy** as heat and sound has a really **low efficiency**, and vice versa.

There's a nice little **equation** to find the **efficiency** of a device:

$$\text{Efficiency} = \frac{\text{useful output energy}}{\text{total input energy}} \times 100\%$$

Energy, as always, is measured in joules (J). Efficiency has no units because it's a ratio.

Some questions will be kind and **give you** the **useful output energy** — others will tell you how much is **wasted**. You just have to **subtract** the **wasted energy** from the **total input energy** to find the **useful output energy**, so it's not too tricky if you keep your wits about you.

Sankey Diagrams Show Energy Input and Output

Sankey diagrams (or energy transformation diagrams) are a way of showing how much of the **input energy** is being **usefully employed** compared with how much is being **wasted**.

For example, the **Sankey diagram** for an **electric motor** is shown below.

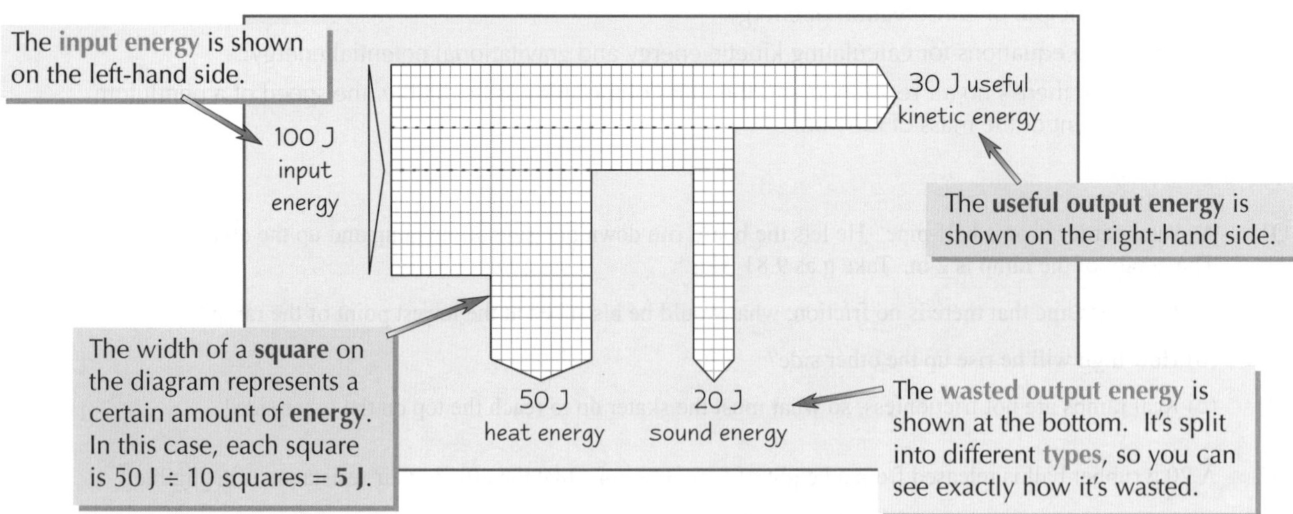
The input energy is shown on the left-hand side.
100 J input energy
30 J useful kinetic energy
The useful output energy is shown on the right-hand side.
The width of a square on the diagram represents a certain amount of energy. In this case, each square is 50 J ÷ 10 squares = 5 J.
50 J heat energy
20 J sound energy
The wasted output energy is shown at the bottom. It's split into different types, so you can see exactly how it's wasted.

The **useful** thing about **Sankey diagrams** is that you can see what's going on at a glance — there's always **one** big **arrow** going in (**input energy**) and a **few** smaller ones **coming out** (**output energy**). The **width** of the arrows tells you how much **energy** is in each form — the **thicker** the arrow, the **greater** the amount of **energy**.

You can even use **Sankey diagrams** to work out the **efficiency** of the device — read off the **total input energy** and **useful output energy**, then substitute them in the **equation** above.

Efficiency and Sankey Diagrams

Drawing Sankey Diagrams is Easy — if You Take it Step by Step

In the exam, you could be asked to **interpret** a Sankey diagram — or you could be asked to **draw** one for yourself. Take it **step by step** and you'll be fine.

1) Find the **total input energy**, the **useful output energy** and the **amount of energy wasted** in each different form. You might be given **all** these values — or you might be given **some** and have to **add** or **subtract** to find the rest.
2) Choose your **scale**. It's best (and easiest to draw) if you can represent **all** the energy values by a **whole number** of squares. It might sound obvious, but you also want your **diagram** to be a **sensible size**.
3) Work out **how many squares** will represent each energy value: **energy ÷ energy per square = number of squares**.
4) **Draw** the **arrow** for the **total input energy** — make sure it's the right number of squares or it'll all go wrong.
5) **Split** the **input energy** into all the different **outputs** — the **useful energy** output should go at the **top**.
6) **Draw** the **output arrows** — **useful energy** should go **straight across**, and **wasted energy** should point **downwards**.
7) **Label** all the **arrows** so it's clear what each bit represents.

Example

A car uses 280 MJ of chemical energy to travel 100 km. During this time, 180 MJ are wasted as heat energy and 30 MJ are wasted as sound energy. Draw a Sankey diagram for the car, then calculate its efficiency.

Start by listing the energy values: **Total input energy = 280 MJ**, **Energy wasted as heat = 180 MJ**, **Energy wasted as sound = 30 MJ**, **Useful output energy** = 280 – 180 – 30 = **70 MJ**.

All the values divide by 10, so use a **scale** of **1 square : 10 MJ**.

Draw the **input arrow**.
280 MJ ÷ 10 = **28 squares**

Split it into the different **outputs**.

Draw the remaining **arrows**, then **label** them all.

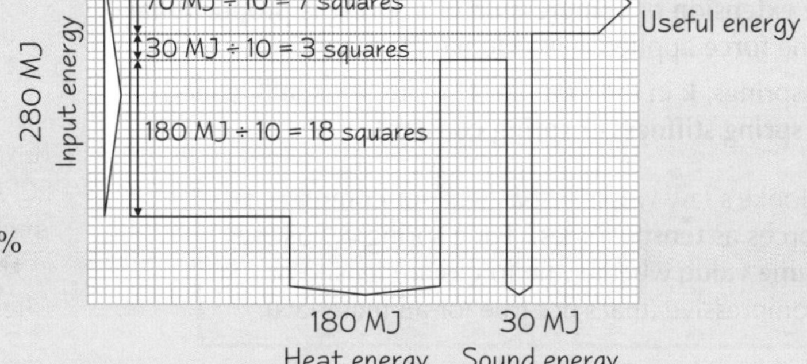

Work out the efficiency:

$$\text{Efficiency} = \frac{\text{useful output energy}}{\text{total input energy}} \times 100\%$$

$$= \frac{70}{280} \times 100 = \mathbf{25\%}$$

Practice Questions

Q1 Why can a device never be 100% efficient?

Q2 What is the equation for efficiency?

Q3 Calculate the efficiency of a device that wastes 65 J for every 140 J of input energy.

Q4 What are Sankey diagrams used for?

Exam Question

Q1 The figure on the right is a Sankey diagram for a certain design of wind turbine. A second design of wind turbine produces 30 kJ of electrical energy for every 125 kJ of input kinetic energy. It wastes 70 kJ as sound, and the rest as heat.

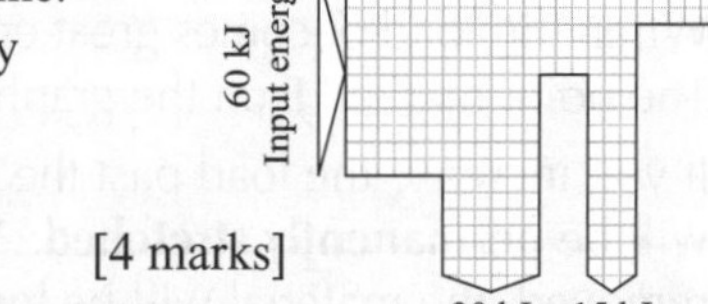

(a) Draw and label a Sankey diagram for the second design of wind turbine. [4 marks]

(b) Which design of wind turbine is more efficient? By how much? [3 marks]

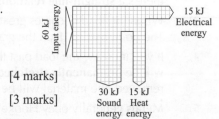

Sankey diagrams are a physicist's best friend...

I'm quite a fan of Sankey diagrams — you take a load of numbers that don't mean very much, draw a nice Sankey diagram, and hey presto, you can see what's going on. Whether you agree or not, make sure you can draw and interpret them.

Hooke's Law

Hooke's law doesn't apply to all materials, and only works for the rest up to a point, but it's still pretty handy.

Hooke's Law Says that Extension is Proportional to Force

If a **metal wire** is supported at the top and then a weight attached to the bottom, it **stretches**. The weight pulls down with force ***F***, producing an equal and opposite force at the support.

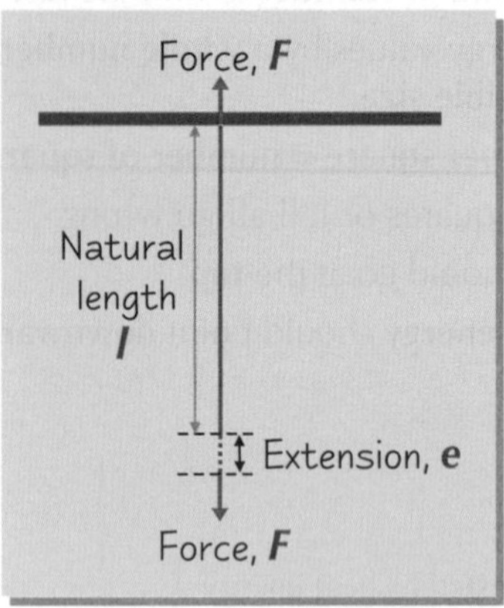

1) **Robert Hooke** discovered in 1676 that the extension of a stretched wire, **e**, is proportional to the load or force, ***F***. This relationship is now called **Hooke's law**.
2) Hooke's law can be written:

$$F = ke$$

Where **k** is a constant that depends on the material being stretched. **k** is called the **stiffness constant**.

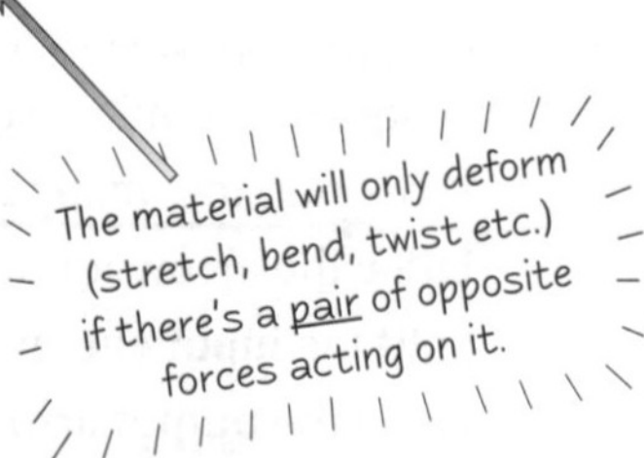

The material will only deform (stretch, bend, twist etc.) if there's a pair of opposite forces acting on it.

I'm a bit irrelevant on this page — bungee ropes don't obey Hooke's Law... Do you think I need to get out more?

Hooke's law Also Applies to Springs

A metal spring also changes length when you apply a **pair of opposite forces**.

1) The **extension** or **compression** of a spring is **proportional** to the **force** applied — so Hooke's law applies.
2) For springs, **k** in the formula ***F* = ke** is usually called the **spring stiffness** or **spring constant**.

Hooke's law works just as well for **compressive** forces as **tensile** forces. For a spring, **k** has the **same value** whether the forces are tensile or compressive (that's not true for all materials).

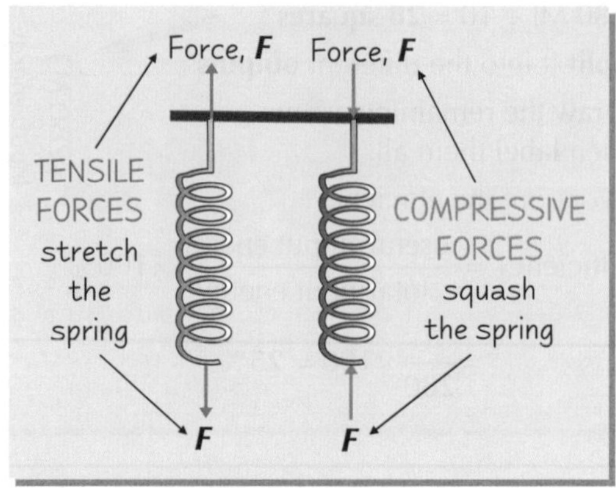

Hooke's law Stops Working when the Load is Great Enough

There's a **limit** to the force you can apply for Hooke's law to stay true.

1) The graph shows load against extension for a **typical metal wire**.
2) The first part of the graph shows Hooke's law being obeyed — there's a **straight-line relationship** between **load** and **extension**.
3) When the load becomes great enough, the graph starts to **curve**. The point marked E on the graph is called the **elastic limit**.
4) If you increase the load past the elastic limit, the material will be **permanently stretched**. When all the force is removed, the material will be **longer** than at the start.
5) **Metals** generally obey Hooke's law up to the limit of proportionality (see p.40), which is very near the elastic limit.
6) Be careful — there are some materials, like **rubber**, that only obey Hooke's law for **really small** extensions.

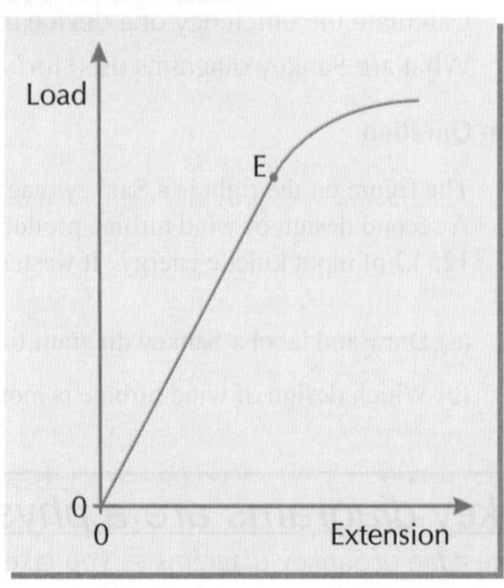

Hooke's Law

So basically...

*A Stretch can be **Elastic** or **Plastic***

Elastic

If a **deformation** is **elastic**, the material returns to its **original shape** once the forces are removed.

1) When the material is put under **tension**, the **atoms** of the material are **pulled apart** from one another.
2) Atoms can **move** small distances relative to their **equilibrium positions**, without actually changing position in the material.
3) Once the **load** is **removed**, the atoms **return** to their **equilibrium** distance apart.

For a metal, elastic deformation happens as long as **Hooke's law** is obeyed.

Plastic

If a deformation is **plastic**, the material is **permanently stretched**.

1) Some atoms in the material move position relative to one another.
2) When the load is removed, the **atoms don't return** to their original positions.

A metal stretched **past its elastic limit** shows plastic deformation.

Practice Questions

Q1 State Hooke's law.

Q2 Define tensile forces and compressive forces.

Q3 Explain what is meant by the elastic limit of a material.

Q4 From studying the force-extension graph for a material as it is loaded and unloaded, how can you tell:
(a) if Hooke's law is being obeyed,
(b) if the elastic limit has been reached?

Q5 What is plastic behaviour of a material under load?

Exam Questions

Q1 A metal guitar string stretches 4.0 mm when a 10 N force is applied.

(a) If the string obeys Hooke's law, how far will the string stretch with a 15 N force? [1 mark]

(b) Calculate the stiffness constant for this string in Nm^{-1}. [2 marks]

(c) The string is tightened beyond its elastic limit. What would be noticed about the string? [1 mark]

Q2 A rubber band is 6.0 cm long. When it is loaded with 2.5 N, its length becomes 10.4 cm. Further loading increases the length to 16.2 cm when the force is 5.0 N.

Does the rubber band obey Hooke's law when the force on it is 5.0 N? Justify your answer with a suitable calculation. [2 marks]

Sod's Law — if you don't learn it, it'll be in the exam...

Three things you didn't know about Robert Hooke — he was the first person to use the word 'cell' (in terms of biology, not prisons), he helped Christopher Wren with his designs for St. Paul's Cathedral and no-one knows quite what he looked like. I'd like to think that if I did all that stuff, then someone would at least remember what I looked like — poor old Hooke.

Stress and Strain

How much a material stretches for a particular applied force depends on its dimensions.
If you want to compare the properties of two different materials, you need to use stress and strain instead.
A stress-strain graph is the same for any sample of a particular material — the size of the sample doesn't matter.

A Stress Causes a Strain

A material subjected to a pair of **opposite forces** might **deform**, i.e. **change shape**. If the forces **stretch** the material, they're **tensile**. If the forces **squash** the material, they're **compressive**.

1) **Tensile stress** is defined as the **force applied**, ***F***, divided by the **cross-sectional area**, ***A***:

$$\text{stress} = \frac{F}{A}$$

The **units** of stress are **Nm^{-2}** or pascals, **Pa**.

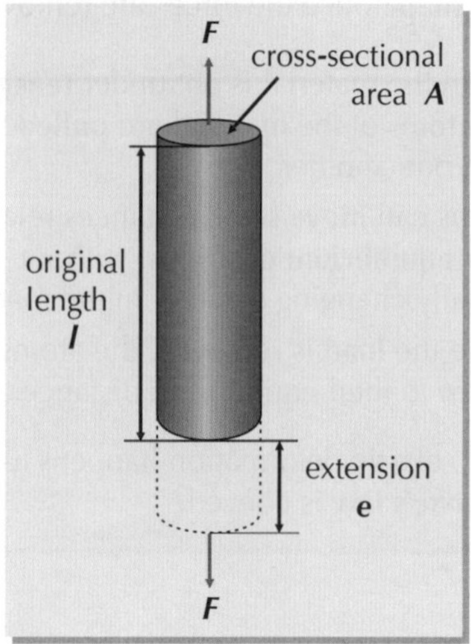

2) **Tensile strain** is defined as the **change in length**, i.e. the **extension**, divided by the **original length** of the material:

$$\text{strain} = \frac{e}{l}$$

Strain has **no units** — it's just a **number**.

3) It doesn't matter whether the forces producing the **stress** and **strain** are **tensile** or **compressive** — the **same equations** apply. The only difference is that you tend to think of **tensile** forces as **positive**, and **compressive** forces as **negative**.

A Stress Big Enough to Break the Material is Called the Breaking Stress

As a greater and greater tensile **force** is applied to a material, the **stress** on it **increases**.

1) The effect of the **stress** is to start to **pull** the **atoms apart** from one another.
2) Eventually the stress becomes **so great** that atoms **separate completely**, and the **material breaks**. This is shown by point **B** on the graph. The stress at which this occurs is called the **breaking stress**.
3) The point marked **UTS** on the graph is called the **ultimate tensile stress**. This is the **maximum stress** that the material can withstand.
4) **Engineers** have to consider the **UTS** and **breaking stress** of materials when designing a **structure**.

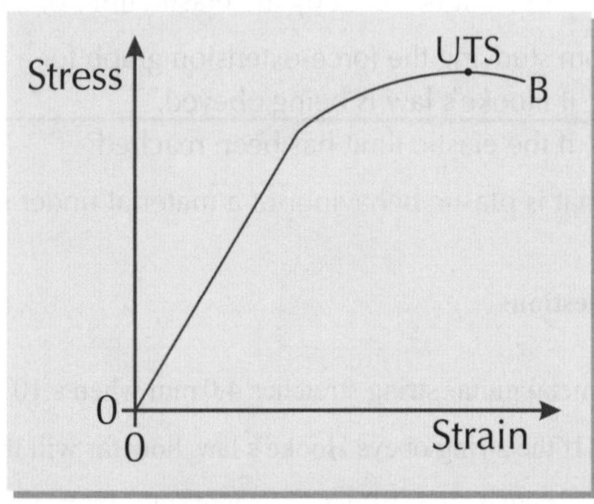

Elastic Strain Energy is the Energy Stored in a Stretched Material

When a material is **stretched**, **work** has to be done in stretching the material.

1) **Before** the **elastic limit**, **all** the **work done** in stretching is **stored** as **potential energy** in the material.
2) This stored energy is called **elastic strain energy**.
3) On a **graph** of **force against extension**, the elastic strain energy is given by the **area under the graph**.

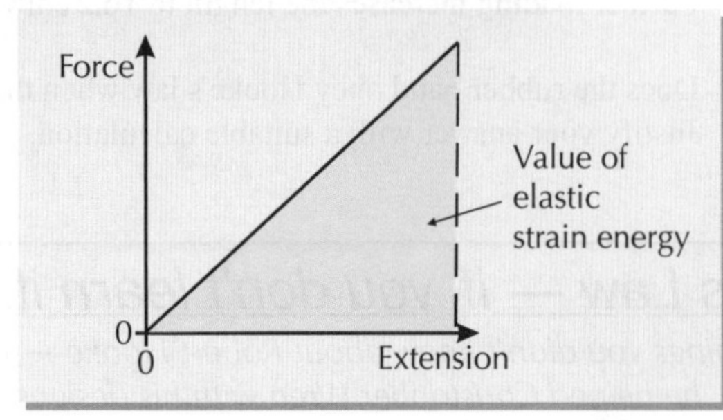

Stress and Strain

You can Calculate the Energy Stored in a Stretched Wire

Provided a material obeys Hooke's law, the **potential energy** stored inside it can be **calculated** quite easily.

1) The work done on the wire in stretching it is equal to the energy stored.
2) **Work done** equals **force × displacement**.
3) However, the **force** on the material **isn't constant**. It rises from zero up to force ***F***. To calculate the **work done**, use the average force between zero and ***F***, i.e. ½***F***.

   $$\text{work done} = \tfrac{1}{2}F \times e$$

4) Then the **elastic strain energy**, ***E***, is:

   $$E = \tfrac{1}{2}Fe$$

   This is the triangular area under the force-extension graph — see previous page.

5) Because Hooke's law is being obeyed, ***F* = *ke***, which means ***F*** can be replaced in the equation to give:

   $$E = \tfrac{1}{2}ke^2$$

6) If the material is stretched beyond the **elastic limit**, some work is done separating atoms. This will **not** be **stored** as strain energy and so isn't available when the force is released.

Practice Questions

Q1 Write a definition for tensile stress.

Q2 Explain what is meant by the tensile strain on a material.

Q3 What is meant by the breaking stress of a material?

Q4 How can the elastic strain energy be found from the force against extension graph of a material under load?

Q5 The work done is usually calculated as force multiplied by displacement. Explain why the work done in stretching a wire is ½*Fe*.

Exam Questions

Q1 A steel wire is 2.00 m long. When a 300 N force is applied to the wire, it stretches 4.0 mm. The wire has a circular cross-section with a diameter of 1.0 mm.

(a) What is the cross-sectional area of the wire? [1 mark]

(b) Calculate the tensile stress in the wire. [1 mark]

(c) Calculate the tensile strain of the wire. [1 mark]

Q2 A copper wire (which obeys Hooke's law) is stretched by 3.0 mm when a force of 50 N is applied.

(a) Calculate the stiffness constant for this wire in Nm^{-1}. [2 marks]

(b) What is the value of the elastic strain energy in the stretched wire? [1 mark]

Q3 A pinball machine contains a spring which is used to fire a small, 12 g metal ball to start the game. The spring has a stiffness constant of 40.8 Nm^{-1}. It is compressed by 5 cm and then released to fire the ball.

Calculate the maximum possible speed of the ball. [4 marks]

UTS a laugh a minute, this stuff...

Here endeth the proper physics for this section — the rest of it's materials science (and I don't care what your exam boards say). It's all a bit "useful" for my liking. Calls itself a physics course... grumble... grumble... wasn't like this in my day... But to be fair — some of it's quite interesting, and there are some pretty pictures coming up on page 66.

The Young Modulus

Busy chap, Thomas Young. He did this work on tensile stress as something of a sideline. Light was his main thing. He proved that light behaved like a wave, explained how we see in colour and worked out what causes astigmatism.

The **Young Modulus** is Stress ÷ Strain

When you apply a **load** to stretch a material, it experiences a **tensile stress** and a **tensile strain**.

1) Up to a point called the **limit of proportionality** (see p.40), the stress and strain of a material are proportional to each other.
2) So below this limit, for a particular material, stress divided by strain is a constant. This constant is called the **Young modulus**, ***E***.

$$E = \frac{\text{tensile stress}}{\text{tensile strain}} = \frac{F/A}{e/l} = \frac{Fl}{eA}$$

Where, ***F*** = force in N, ***A*** = cross-sectional area in m², ***l*** = initial length in m and **e** = extension in m.

3) The **units** of the Young modulus are the same as stress (**Nm⁻²** or pascals), since strain has no units.
4) The Young modulus is used by **engineers** to make sure their materials can withstand sufficient forces.

To **Find** the Young Modulus, You need a **Very Long Wire**

This is the experiment you're most likely to do in class:

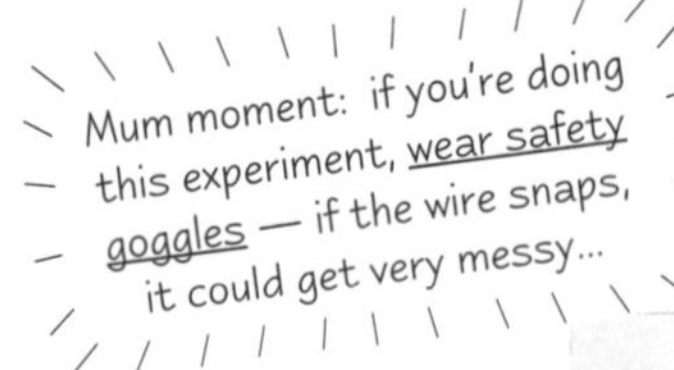

Mum moment: if you're doing
this experiment, wear safety
goggles — if the wire snaps,
it could get very messy...

The Young Modulus

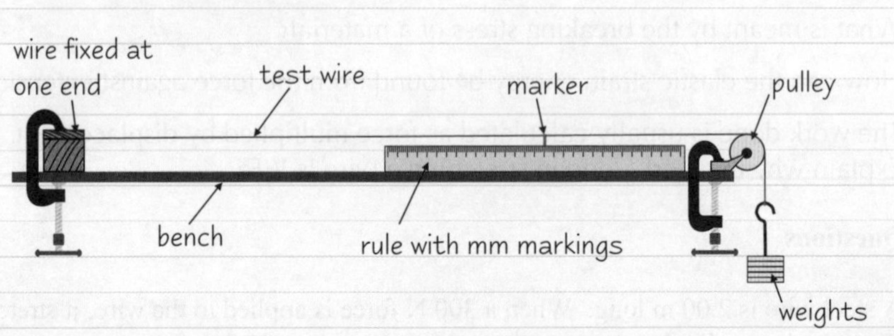

The test wire should be thin, and as long as possible. The **longer and thinner** the wire, the more it **extends** for the same force.

Start with the **smallest weight** necessary to straighten the wire.

Measure the **distance** between the **fixed end of the wire** and the **marker** — this is your unstretched length.

If you then increase the weight, the **wire stretches** and the **marker moves**.

Increase the **weight** by steps, recording the marker reading each time — the **extension** is the **difference** between this reading and the **unstretched length**.

Once you've taken all your readings, use a **micrometer** to measure the **diameter** of the wire in several places. Take an average of your measurements, and use that to work out the average **cross-sectional area** of the wire.

The other standard way of measuring the Young modulus in the lab is using **Searle's apparatus**. This is a bit more accurate, but it's harder to do and the equipment's more complicated.

The Young Modulus

Use a Stress-Strain Graph to Find E

You can plot a **graph** of **stress against strain** from your results.

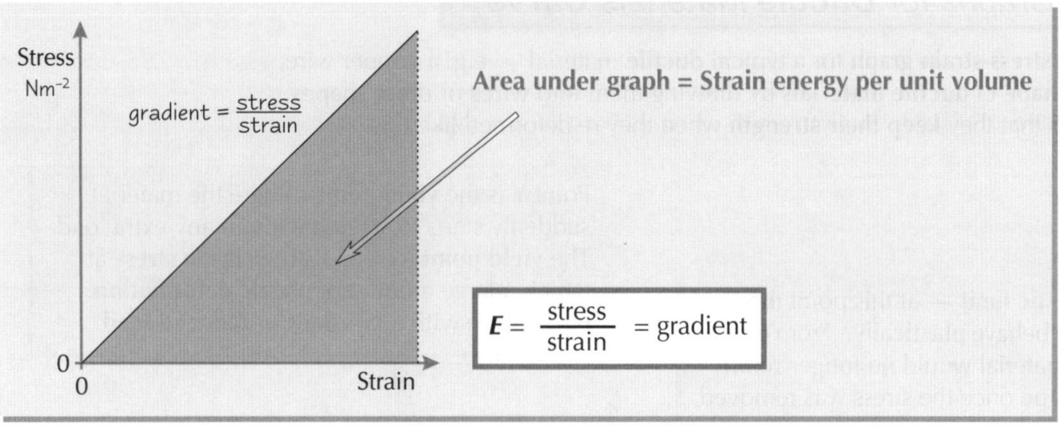

1) The **gradient** of the graph gives the Young modulus, **E**.
2) The **area under the graph** gives the **strain energy** (or energy stored) per unit volume i.e. the energy stored per 1 m^3 of wire.
3) The stress-strain graph is a **straight line** provided that Hooke's law is obeyed, so you can also calculate the energy per unit volume as:

energy = ½ × stress × strain

Practice Questions

Q1 Define the Young modulus for a material.

Q2 What are the units for the Young modulus?

Q3 Explain why a thin test wire is used to find the Young modulus.

Q4 What is given by the area contained under a stress-strain graph?

Exam Questions

Q1 A steel wire is stretched elastically. For a load of 80 N, the wire extends by 3.6 mm. The original length of the wire was 2.50 m and its average diameter is 0.6 mm.

(a) Calculate the cross-sectional area of the wire in m^2. [1 mark]

(b) Find the tensile stress applied to the wire. [1 mark]

(c) Calculate the tensile strain of the wire. [1 mark]

(d) What is the value of the Young modulus for steel? [1 mark]

Q2 The Young modulus for copper is 1.3×10^{11} Nm^{-2}.

(a) If the stress on a copper wire is 2.6×10^{8} Nm^{-2}, what is the strain? [2 marks]

(b) If the load applied to the copper wire is 100 N, what is the cross-sectional area of the wire? [1 mark]

(c) Calculate the strain energy per unit volume for this loaded wire. [1 mark]

Learn that experiment — it's important...

Getting back to the good Dr Young... As if ground-breaking work in light, the physics of vision and materials science wasn't enough, he was also a well-respected physician, a linguist and an Egyptologist. He was one of the first to try to decipher the Rosetta stone (he didn't get it right, but nobody's perfect). Makes you feel kind of inferior, doesn't it. Best get learning.

Interpreting Stress-Strain Graphs

Remember that lovely stress-strain graph from page 36? Well, turns out that because materials have different properties, their stress-strain graphs look different too — you need to know the graphs for ductile, brittle and polymeric materials.

Stress-Strain *Graphs for* ***Ductile*** *Materials* ***Curve***

The diagram shows a **stress-strain graph** for a typical **ductile** material — e.g. a copper wire. You can change the **shape** of **ductile materials** by drawing them into **wires** or other shapes. The important thing is that they **keep their strength** when they're deformed like this.

Point **Y** is the **yield point** — here the material suddenly starts to **stretch** without any extra load. The **yield point** (or yield stress) is the **stress** at which a large amount of **plastic deformation** takes place with a **constant** or **reduced load.**

Point **E** is the **elastic limit** — at this point the material starts to behave **plastically**. From point E onwards, the material would **no longer** return to its **original shape** once the stress was removed.

Point **P** is the **limit of proportionality** — after this, the graph is no longer a straight line but starts to **bend**. At this point, the material **stops** obeying **Hooke's law**, but would still **return** to its **original shape** if the stress was removed.

Before point **P**, the graph is a **straight line** through the **origin**. This shows that the material is obeying **Hooke's law** (page 34).

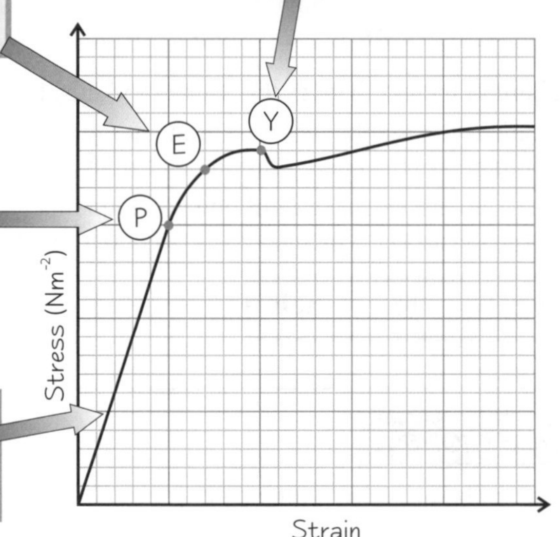

Stress-Strain *Graphs for* ***Brittle*** *Materials* ***Don't Curve***

The graph shown below is typical of a **brittle** material.

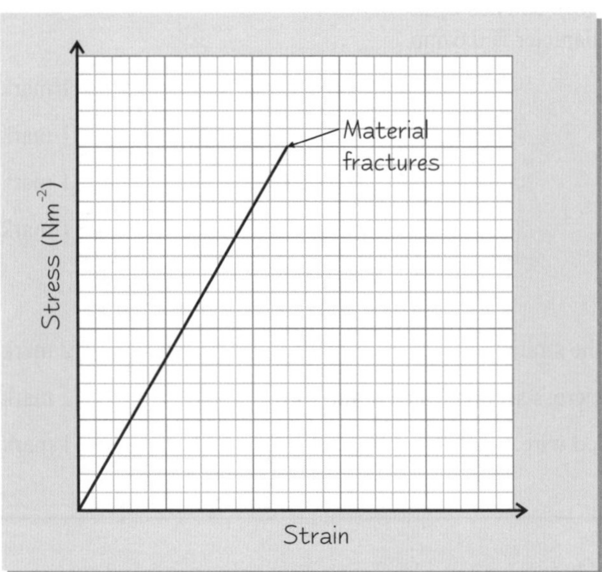

The graph starts the same as the one above — with a **straight line through the origin**. So brittle materials also obey **Hooke's law**.

However, when the **stress** reaches a certain point, the material **snaps** — it doesn't deform plastically.

When **stress** is applied to a brittle material any **tiny cracks** at the material's surface get **bigger** and **bigger** until the material **breaks** completely. This is called **brittle fracture**.

Hooke's law — it's the pirates' code... yarr

Interpreting Stress-Strain Graphs

Rubber and *Polythene* Are *Polymeric Materials*

1) The **molecules** that make up **polymeric** (or polymer) **materials** are arranged in **long chains**.
2) They have a **range** of properties, so different polymers have different **stress-strain graphs**. The diagram below shows two examples — **rubber** and **polythene**.

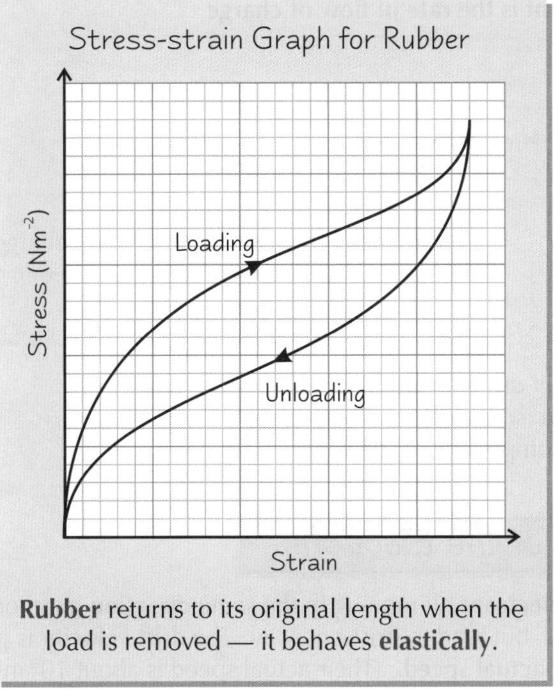

Rubber returns to its original length when the load is removed — it behaves **elastically**.

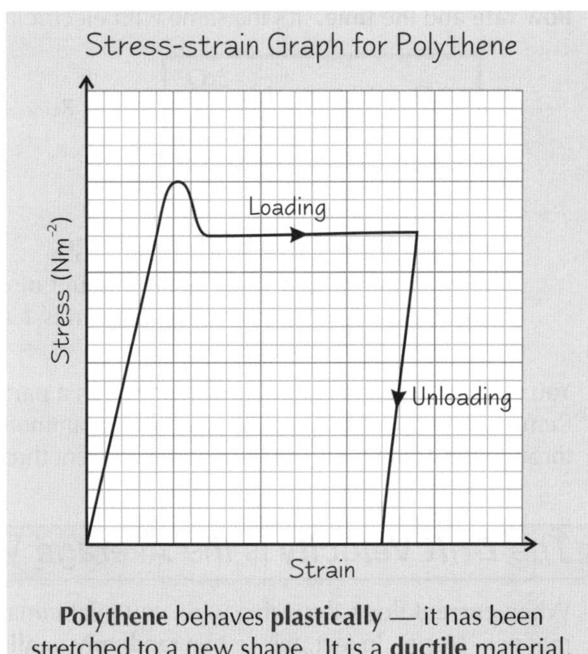

Polythene behaves **plastically** — it has been stretched to a new shape. It is a **ductile** material.

Practice Questions

Q1 Write short definitions of the following terms: ductile, brittle.

Q2 What is the difference between the limit of proportionality and the elastic limit?

Q3 What are polymeric materials?

Q4 Sketch stress-strain graphs of typical ductile, brittle and polymeric materials and describe their shapes.

Exam Questions

Q1 Hardened steel is a hard, brittle form of steel made by heating it up slowly and then quenching it in cold water.

(a) What is meant by the term *brittle*? [2 marks]

(b) Sketch a stress-strain graph for hardened steel. [2 marks]

Q2 An electric cable consists of a copper wire surrounded by polythene insulation.

(a) Sketch stress-strain graphs for the two materials. [3 marks]

(b) Describe two similarities between the behaviour of the two materials under stress. [2 marks]

My sister must be brittle — she's always snapping...

In case you were wondering, I haven't just drawn the graphs on these two pages for fun (though I did enjoy myself) — they're there for you to learn. I find the best way to remember each one is to understand why it has the shape it does — if that sounds too much like hard work, then at least make sure you can describe the shape of all four of them.

Charge, Current and Potential Difference

You wouldn't reckon there was that much to know about electricity... just plug something in, and bosh — electricity. Ah well, never mind the age of innocence — here are all the gory details...

Current is the Rate of Flow of Charge

The **current** in a **wire** is like **water** flowing in a **pipe**. The **amount** of water that flows depends on the **flow rate** and the **time**. It's the same with electricity — **current is the rate of flow of charge**.

$$\Delta Q = I\Delta t \quad \text{or} \quad I = \frac{\Delta Q}{\Delta t}$$

Where ΔQ is the charge in coulombs, I is the current and Δt is the time taken.

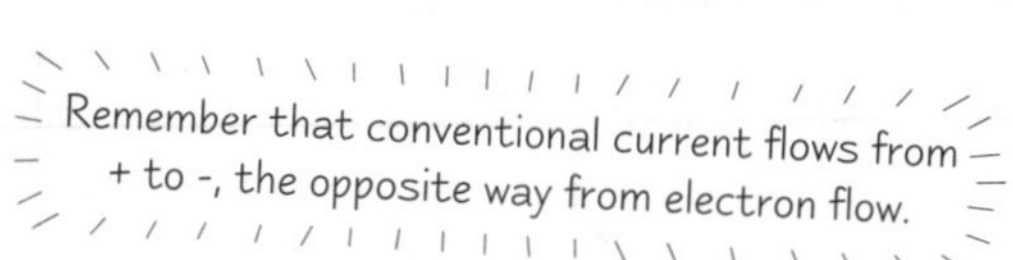

Remember that conventional current flows from + to -, the opposite way from electron flow.

The Coulomb is the Unit of Charge

One **coulomb** (**C**) is defined as the **amount of charge** that passes in **1 second** when the **current** is **1 ampere**.

You need to know the elementary charge too (i.e. the charge on a single electron):

$e = 1.6 \times 10^{-19}$ C

You can measure the current flowing through a part of a circuit using an **ammeter**. Remember — you always need to attach an ammeter in **series** (so that the current through the ammeter is the same as the current through the component — see page 54).

The Drift Velocity is the Average Velocity of the Electrons

When **current** flows through a wire, you might imagine the **electrons** all moving in the **same direction** in an orderly manner. Nope. In fact, they move **randomly** in **all directions**, but tend to **drift** one way. The **drift velocity** is just the **average velocity** and it's **much, much less** than the electrons' **actual speed**. (Their actual speed is about 10^6 ms^{-1}!)

The Current Depends on the Drift Velocity

The **current** is given by the equation: $I = nAvq$

You don't need to derive this for the exam but you do need to understand what it means.

where: I = electric current in A
n = number of charge carriers per m^3
A = cross-sectional area in m^2
v = drift velocity in ms^{-1}
q = charge in C carried by each charge carrier

See what the Equation Means by Changing One Variable at a Time

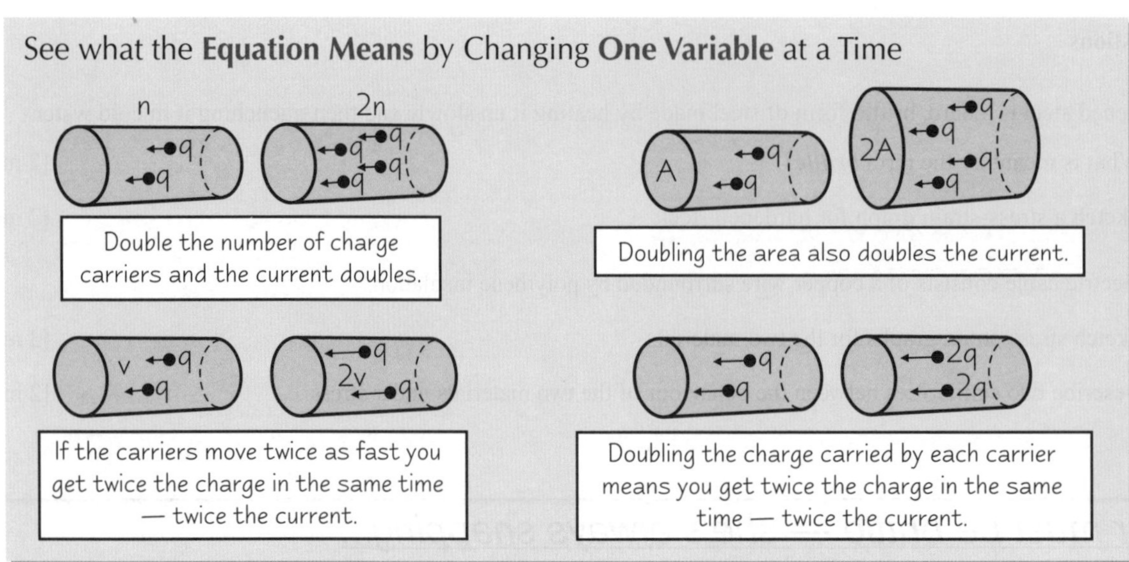

Double the number of charge carriers and the current doubles.

Doubling the area also doubles the current.

If the carriers move twice as fast you get twice the charge in the same time — twice the current.

Doubling the charge carried by each carrier means you get twice the charge in the same time — twice the current.

Charge, Current and Potential Difference

Different Materials** have Different Numbers of **Charge Carriers

1) In a **metal**, the **charge carriers** are **free electrons** — they're the ones from the **outer shell** of each atom. Thinking about the formula $I = nAvq$, there are **loads** of charge carriers, making n **big**. The **drift velocity** only needs to be **small**, even for a **high current**.
2) **Semiconductors** have **fewer charge carriers** than metals, so the **drift velocity** will need to be **higher** if you're going to have the **same current**.
3) A **perfect insulator** wouldn't have **any charge carriers**, so $n = 0$ in the formula and you'd get **no current**. **Real** insulators have a **very small** n.

Charge Carriers** in Liquids and Gases are **Ions

1) **Ionic crystals** like sodium chloride are **insulators**. Once **molten**, though, the liquid **conducts**. Positive and negative **ions** are the **charge carriers**. The **same thing** happens in an **ionic solution** like copper sulfate solution.
2) **Gases** are **insulators**, but if you apply a **high enough voltage** electrons get **ripped out** of **atoms**, giving you **ions** along a path. You get a **spark**.

Potential Difference** is the **Energy** per **Unit Charge

To make electric charge flow through a conductor, you need to do work on it. **Potential difference** (p.d.), or **voltage**, is defined as the **energy converted per unit charge moved**.

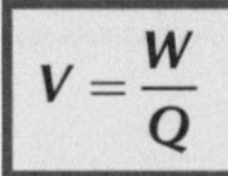

$$V = \frac{W}{Q}$$

W is the energy in joules. It's the work you do moving the charge.

Back to the 'water analogy' again. The p.d. is like the pressure that's forcing water along the pipe.

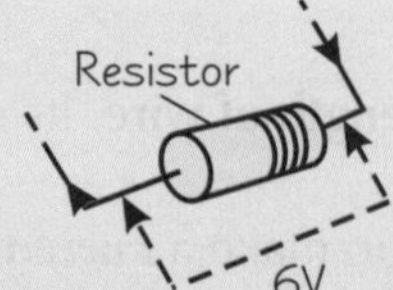

Here you do 6 J of work moving each coulomb of charge through the resistor, so the p.d. across it is 6 V. The energy gets converted to heat.

Definition of the **Volt**

The **potential difference** across a component is **1 volt** when you convert **1 joule** of energy moving **1 coulomb** of charge through the component.

$$1\ \text{V} = 1\ \text{J C}^{-1}$$

Practice Questions

Q1 Describe in words how current and charge are related.

Q2 Define the coulomb.

Q3 Explain what drift velocity is.

Q4 Define potential difference.

Exam Questions

Q1 A battery delivers 4500 C of electric charge to a circuit in 10 minutes. Calculate the average current. [2 marks]

Q2 Copper has 1.0×10^{29} free electrons per m^3. Calculate the drift velocity of the electrons in a copper wire of cross-sectional area $5.0 \times 10^{-6}\ m^2$ when it is carrying a current of 13 A. (electron charge = 1.6×10^{-19} C) [3 marks]

Q3 An electric motor runs off a 12 V d.c. supply and has an overall efficiency of 75%. Calculate how much electric charge will pass through the motor when it does 90 J of work. [3 marks]

I can't even be bothered to make the current joke...

Talking of currant jokes, I saw this bottle of wine the other day called 'raisin d'être' — 'raison d'être' of course meaning 'reason for living', but spelled slightly different to make 'raisin', meaning 'grape'. Ho ho. Chuckled all the way out of Tesco.

Resistance and Resistivity

"You will be assimilated. Resistance is futile."
Sorry, I couldn't resist it (no pun intended), and I couldn't think of anything useful to write anyway. This resistivity stuff gets a bit more interesting when you start thinking about temperature and light dependence, but for now, just learn this.

Everything has Resistance

1) If you put a **potential difference** (p.d.) across an **electrical component**, a **current** will flow.
2) **How much** current you get for a particular **p.d.** depends on the **resistance** of the component.
3) You can think of a component's **resistance** as a **measure** of how **difficult** it is to get a **current** to **flow** through it.

Mathematically, **resistance** is: $$R = \frac{V}{I}$$

This equation really **defines** what is meant by resistance.

4) **Resistance** is measured in **ohms** (Ω).

A component has a resistance of **1 Ω** if a **potential difference** of **1 V** makes a **current** of **1 A** flow through it.

Three Things Determine Resistance

If you think about a nice, **simple electrical component**, like a **length of wire**, its **resistance** depends on:

1) **Length** (l). The **longer** the wire the **more difficult** it is to make a **current flow**.
2) **Area** (A). The **wider** the wire the **easier** it will be for the electrons to pass along it.
3) **Resistivity** (ρ). This **depends** on the **material**. The **structure** of the material of the wire may make it easy or difficult for charge to flow. In general, resistivity depends on **environmental factors** as well, like **temperature** and **light intensity**.

The **resistivity** of a material is defined as the **resistance** of a **1m length** with a **1m² cross-sectional area**. It is measured in **ohm metres** (Ωm).

This is the Greek letter rho, the symbol for resistivity.

$$\rho = \frac{RA}{l}$$

where A = cross-sectional area in m², and l = length in m

You will more **usually** see the equation in the **form**: $$R = \rho\frac{l}{A}$$

Typical values for the **resistivity** of **conductors** are **really small**.
For example, the resistivity of **copper** (at 25 °C) is just 1.72×10^{-8} Ωm.

If you **calculate** a **resistance** for a **conductor** and end up with something **really small** (e.g. 1×10^{-7} Ω), go back and **check** that you've **converted** your **area** into **m²**.
It's really easy to make mistakes with this equation by leaving the area in **cm²** or **mm²**.

Resistance and Resistivity

For an Ohmic Conductor, R is a Constant

A chap called **Ohm** did most of the early work on resistance. He developed a rule to **predict** how the **current** would **change** as the applied **potential difference increased**, for **certain types** of conductor.

The rule is now called **Ohm's law** and the conductors that **obey** it (mostly metals) are called **ohmic conductors**.

Provided the **temperature** is **constant**, the **current** through an ohmic conductor is **directly proportional** to the **potential difference** across it.

$$R = \frac{V}{I}$$

1) As you can see from the graph, **doubling** the **p.d. doubles** the **current**.
2) What this means is that the **resistance** is **constant**.
3) Often **external factors**, such as **light level** or **temperature** will have a **significant effect** on resistance, so you need to remember that Ohm's law is **only** true for **ohmic conductors** at **constant temperature**.

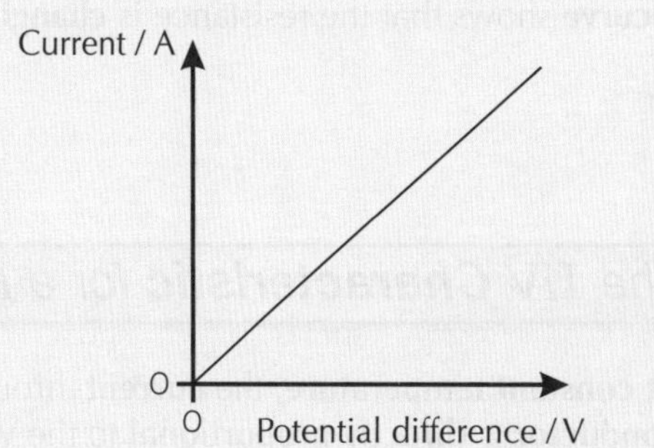

Practice Questions

Q1 Name one environmental factor likely to alter the resistance of a component.

Q2 What is special about an ohmic conductor?

Q3 What three factors does the resistance of a length of wire depend on?

Q4 What are the units for resistivity?

Exam Questions

Q1 Aluminium has a resistivity of $2.8 \times 10^{-8}\ \Omega$ m at 20 °C.

Calculate the resistance of a pure aluminium wire of length 4 m and diameter 1 mm, at 20 °C. [3 marks]

Q2 The table below shows some measurements taken by a student during an experiment investigating an unknown electrical component.

Potential Difference (V)	Current (mA)
2.0	2.67
7.0	9.33
11.0	14.67

(a) Use the first row of the table to calculate the resistance of the component when a p.d. of 2 V is applied. [2 marks]

(b) By means of further calculation, or otherwise, decide whether the component is an ohmic conductor. [3 marks]

Resistance and resistivity are NOT the same...

Superconductors are great. You can use a magnet to set a current flowing in a loop of superconducting wire. Take away the magnet and the current keeps flowing forever. A wire with no resistance never loses any energy. Pretty cool, huh.

Woohoo — real physics. This stuff's actually kind of interesting.

I/*V* Graphs Show how Resistance Varies

The term '***I*/*V* characteristic**' refers to a **graph** which shows how the **current** (***I***) flowing through a **component changes** as the **potential difference** (***V***) across it is increased.

The **shallower** the **gradient** of a characteristic ***I*/*V*** graph, the **greater** the **resistance** of the component.

A **curve** shows that the resistance is **changing**.

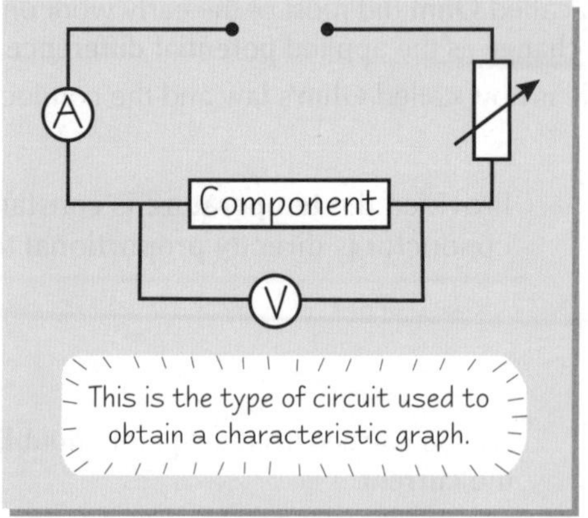

A
Component
V
This is the type of circuit used to obtain a characteristic graph.

The *I*/*V* Characteristic for a Metallic Conductor is a Straight Line

At **constant temperature**, the **current** through a **metallic conductor** is **directly proportional** to the **voltage**.
The fact that the characteristic graph is a **straight line** tells you that the **resistance doesn't change**. **Metallic conductors** are **ohmic** — they have **constant resistance provided** their temperature doesn't change.

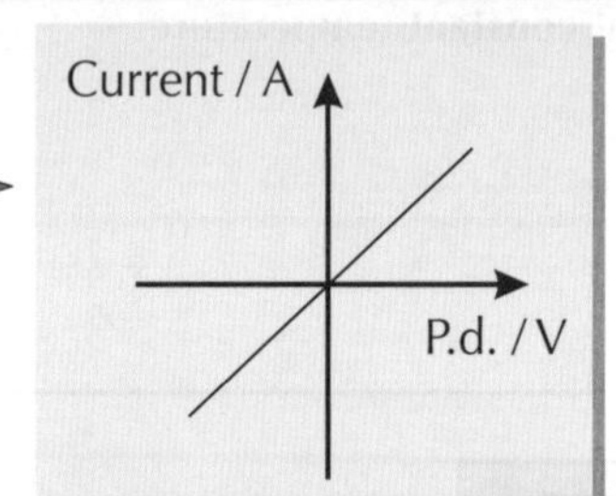

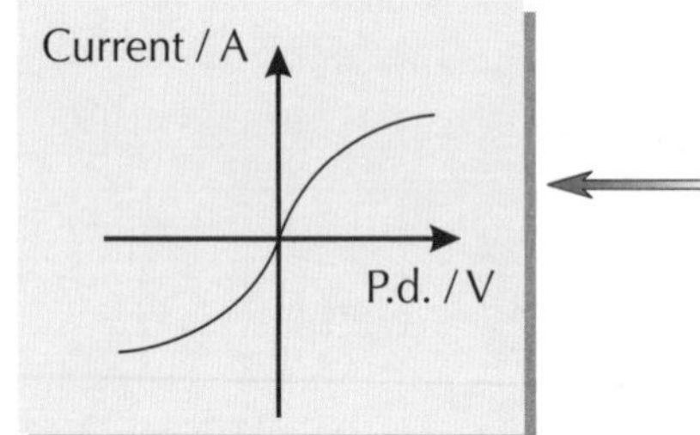

The characteristic graph for a **filament lamp** is a **curve**, which starts **steep** but gets **shallower** as the **voltage rises**. The **filament** in a lamp is just a **coiled up** length of **metal wire**, so you might think it should have the **same characteristic graph** as a **metallic conductor**. It doesn't because it **gets hot**. **Current** flowing through the lamp **increases** its **temperature**.

The **resistance** of a **metal increases** as the **temperature increases**.

The Temperature Affects the Charge Carriers

1) **Charge** is carried through **metals** by **free electrons** in a **lattice** of **positive ions**.
2) Heating up a metal doesn't affect how many electrons there are, but it does make it **harder** for them to **move about**. The **ions vibrate more** when heated, so the electrons **collide** with them more often, **losing energy**.

The **resistance** of most metallic conductors **goes up linearly** with **temperature**.

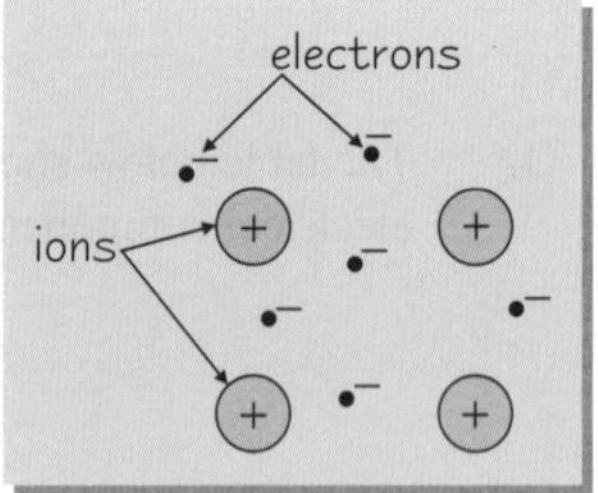

Semiconductors are Used in Sensors

Semiconductors are **nowhere near** as good at **conducting** electricity as **metals**. This is because there are far, far **fewer charge carriers** available. However, if **energy** is supplied to the semiconductor, **more charge carriers** are often **released**. This means that they make **excellent sensors** for detecting **changes** in their **environment**.

You need to know about **three** semiconductor components — **thermistors**, **LDRs** and **diodes**.

I/V Characteristics

The *Resistance* of a *Thermistor* Depends on *Temperature*

Thermistor circuit symbol:

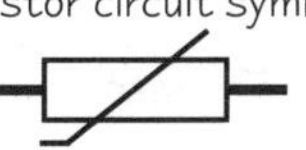

A **thermistor** is a **resistor** with a **resistance** that depends on its **temperature**. You only need to know about **NTC** thermistors — NTC stands for 'Negative Temperature Coefficient'. This means that the **resistance decreases** as the **temperature goes up**. The characteristic graph for an NTC thermistor curves upwards.

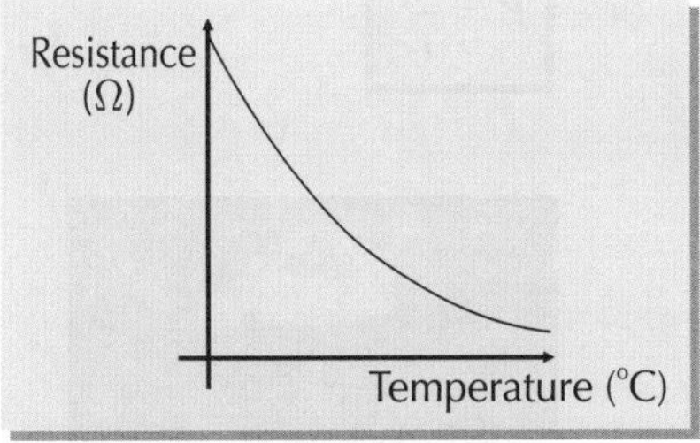

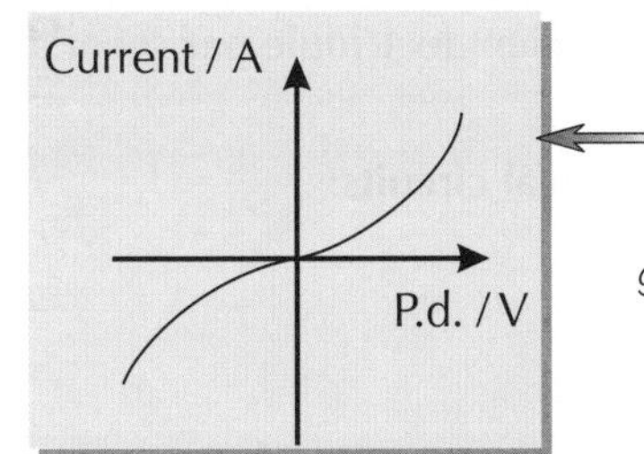

Increasing the current through the thermistor increases its temperature. The increasing gradient of this characteristic graph tells you that the resistance is decreasing.

Warming the thermistor gives more **electrons** enough **energy** to **escape** from their atoms. This means that there are **more charge carriers** available, so the resistance is lower.

The Resistance of an *LDR* depends on *Light Intensity*

LDR circuit symbol:

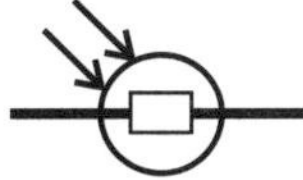

LDR stands for **Light-Dependent Resistor**. The **greater** the intensity of **light** shining on an LDR, the **lower** its **resistance**.

The explanation for this is similar to that for the thermistor. In this case, **light** provides the **energy** that releases more electrons. More charge carriers means a lower resistance.

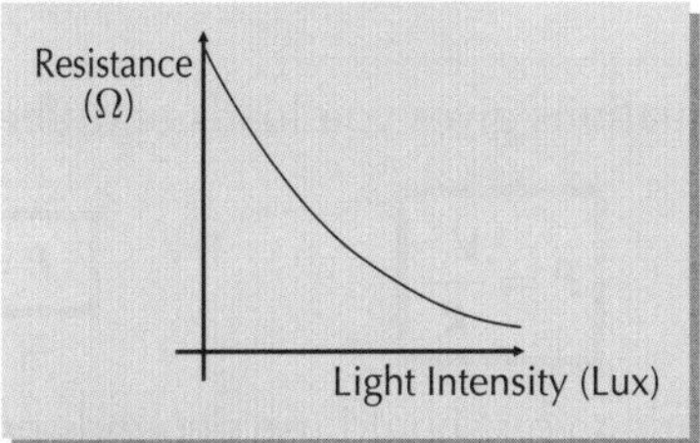

Large Dayglow Rabbit

Diodes Only Let *Current Flow* in *One Direction*

Current / A

P.d. / V

Diode and LED circuit symbols:

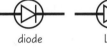

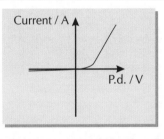

Diodes (including light emitting diodes (LEDs)) are designed to let **current flow** in **one direction** only. You don't need to be able to explain how they work, just what they do.

1) **Forward bias** is the **direction** in which the **current** is **allowed to flow**.
2) **Most** diodes require a **threshold voltage** of about **0.6 V** in the **forward direction** before they will conduct.
3) In **reverse bias**, the **resistance** of the diode is **very high** and the current that flows is **very tiny**.

Practice Questions

Q1 Sketch the circuit used to determine the ***I/V*** characteristics of a component.

Q2 Draw an ***I/V*** characteristic graph for a diode.

Q3 What is an LDR?

Q4 If an ***I/V*** graph is curved, what does this tell you about the resistance?

Exam Question

Q1 (a) Sketch a characteristic ***I/V*** graph for a filament lamp. [1 mark]

(b) State how the resistance changes as the temperature increases. [1 mark]

(c) Explain why this happens. [2 marks]

Thermistor man — temperature-dependent Mr Man...

Learn the graphs on this page, and make sure you can explain them. Whether it's a light-dependent resistor or a thermistor, the same principle applies. More energy releases more charge carriers, and more charge carriers means a lower resistance.

Power and energy are pretty familiar concepts — and here they are again. Same principles, just different equations.

Power is the Rate of Transfer of Energy

Power (P) is **defined** as the **rate** of **transfer** of **energy**.
It's measured in **watts** (W), where **1 watt** is equivalent to **1 joule per second**.

or $P = \frac{E}{t}$

There's a really simple formula for **power** in **electrical circuits**:

$P = VI$

This makes sense, since:

1) **Potential difference** (V) is defined as the **energy transferred** per **coulomb**.
2) **Current** (I) is defined as the **number** of **coulombs** transferred per **second**.
3) So **p.d.** × **current** is **energy transferred per second**, i.e. **power**.

He didn't know when, he didn't know where... but one day this PEt would get his revenge.

You know from the definition of **resistance** that: $V = IR$

Combining the **two equations** gives you loads of **different ways** to **calculate power**.

$P = VI$ $\quad$ $P = \frac{V^2}{R}$ $\quad$ $P = I^2R$

Obviously, which equation you should use depends on what **quantities** you're given in the **question**.

Phew... that's quite a few equations to learn and love. And as if they're not exciting enough, here's some examples to get your teeth into...

Example 1

A 24 W car head lamp is connected to a 12 V car battery.
(a) How much energy will the lamp convert into light and heat energy in 2 hours?
(b) Find the total resistance of the lamp and the wires connecting it to the battery.

(a) Number of seconds in 2 hours = 120 × 60 = 7200 s
$E = P \times t = 24 \times 7200 = 172\,800$ J = **172.8 kJ**

(b) Rearrange the equation $P = \frac{V^2}{R}$, $R = \frac{V^2}{P} = \frac{12^2}{24} = \frac{144}{24} = \underline{6\,\Omega}$

Example 2

A robotic mutant Santa from the future converts 750 J of electrical energy into heat every second.
(a) What is the power rating of the robotic mutant Santa?
(b) All of the robotic mutant Santa's components are connected in series, with a total resistance of 30 Ω. What current flows through his wire veins?

(a) Power (W) = $E \div t = 750 \div 1 =$ **750 W**

(b) Rearrange the equation $P = I^2R$, $I = \sqrt{\frac{P}{R}} = \sqrt{\frac{750}{30}} = \sqrt{25} = \underline{5\text{ A}}$

Electrical Energy and Power

Energy *is Easy to* ***Calculate*** *if you Know the* ***Power***

Sometimes it's the **total energy** transferred that you're interested in. In this case you simply need to **multiply** the **power** by the **time**. So:

$$\boxed{E = VIt} \qquad (\text{or } E = \frac{V^2}{R}t \quad \text{or } E = I^2Rt)$$

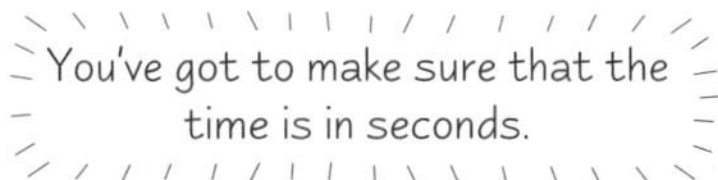

You've got to make sure that the time is in seconds.

Example

Betty pops the kettle on to make a brew.
It takes 4.5 minutes for the kettle to boil the water inside it.
A current of 4 A flows through the kettle's heating element once it is connected to the mains (230 V).

(a) What is the power rating of the kettle?

(b) How much energy does the kettle's heating element transfer to the water in the time it takes to boil?

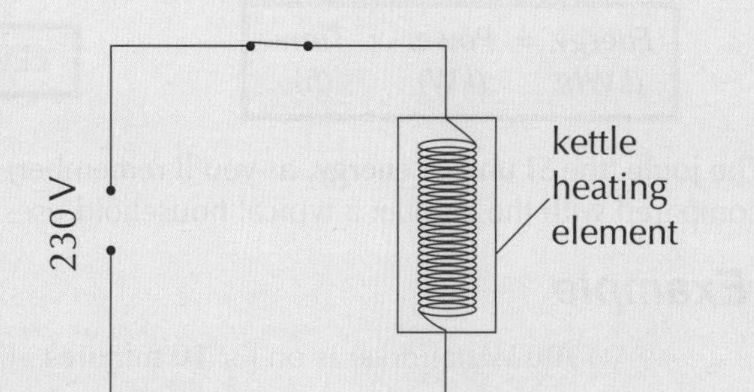

(a) Use $P = V \times I = 230 \times 4 =$ **920 W**

(b) Time the kettle takes to boil in seconds = 4.5 × 60 = 270 seconds.
Use the equation $E = Pt = VIt = 230 \times 4 \times 270 = 248\,400$ J = **248.4 kJ**

Practice Questions

Q1 Write down the equation linking power, current and resistance.

Q2 What equation links power, voltage and resistance?

Q3 Power is measured in watts. What is 1 watt equivalent to?

Exam Questions

Q1 This question concerns a mains powered hairdryer, the circuit diagram for which is given below.

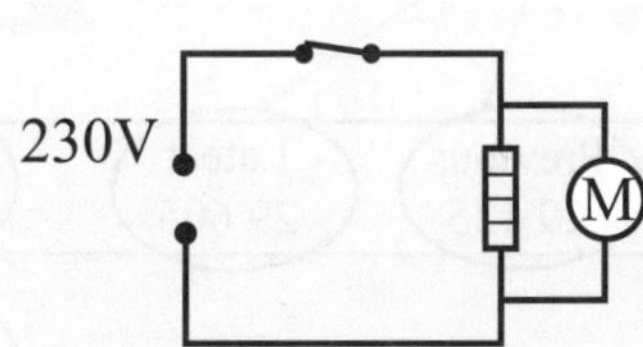

(a) The heater has a power of 920 W in normal operation. Calculate the current in the heater. [2 marks]

(b) The motor has a resistance of 190 Ω. What current will flow in the motor when the hairdryer is used? [2 marks]

(c) Show that the total power of the hairdryer in normal operation is just under 1.2 kW. [2 marks]

Q2 A 12 V car battery supplies a current of 48 A for 2 seconds to the car's starter motor.
The total resistance of the connecting wires is 0.01 Ω.

(a) Calculate the energy transferred from the battery. [1 mark]

(b) Calculate the energy wasted as heat in the wires. [2 marks]

Ultimate cosmic powers...

Whenever you get equations in this book, you know you're gonna have to learn them. Fact of life. I used to find it helped to stick big lists of equations all over my walls in the run up to the exams. But as that's possibly the least cool wallpaper imaginable, I don't advise inviting your friends round till after the exams...

If you went into an electricity shop and asked for a 100 joule packet of electricity you'd be laughed out of town. Why — because electricity companies use units, not joules — phew, you kids don't know anything these days.

Electricity Companies *don't use* **Joules** *and* **Watts**

Electricity companies charge their customers for '**units**' of electricity. Another name for a unit is a **kilowatt-hour (kWh)**. If you know the **power** of an **appliance** and the **length of time** it's used for you can work out the **energy** it uses in kWh.

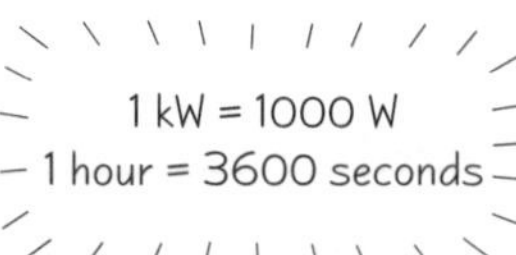

$$\text{Energy (kWh)} = \text{Power (kW)} \times \text{Time (h)}$$

1 kWh = 3.6 million joules

The **joule** (the **SI unit** of **energy**, as you'll remember) is such a **small amount** of energy compared with the amount a typical household uses every month that it's **impractical**.

Example

A **1500 W** hairdryer is on for **10 minutes**. How much energy does it use in J and kWh?

$E = Pt = 1500 \times 10 \times 60 = 900\,000$ J

$E = Pt = 1.5 \times 1/6 = 0.25$ kWh

Cost of Electricity *is the* **Price per Unit** *Times the* **Number of Units** *Used*

Electricity bills can look like they're written in a strange code — but luckily for you, the examples you'll see at AS are easy to understand. Real ones aren't really that bad either — you just need to know **where to look** to find the **important information**. Take a look at the lovely **example** below:

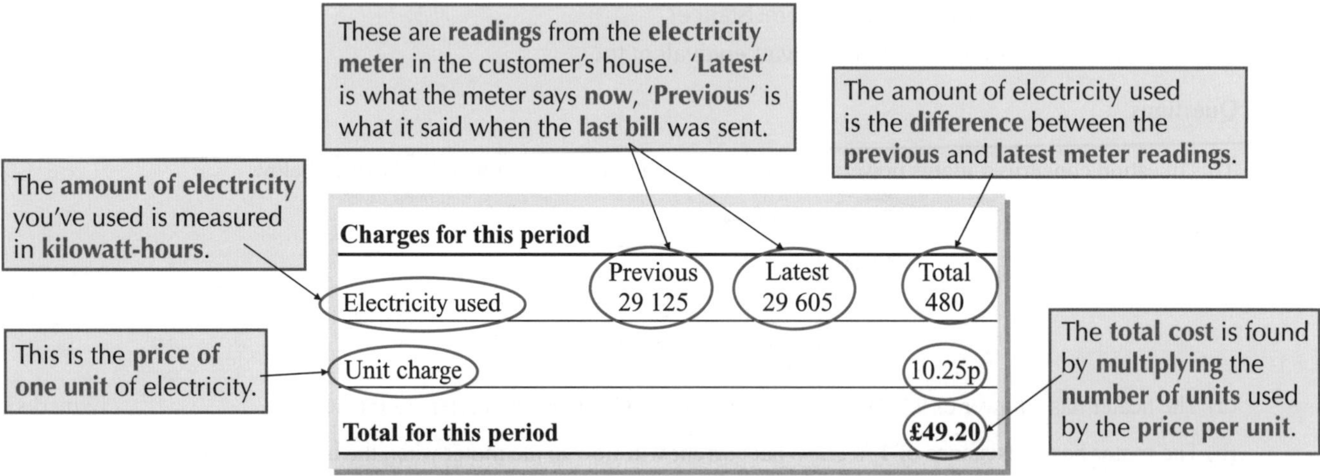

These are readings from the electricity meter in the customer's house. 'Latest' is what the meter says now, 'Previous' is what it said when the last bill was sent.
The amount of electricity used is the difference between the previous and latest meter readings.
The amount of electricity you've used is measured in kilowatt-hours.
This is the price of one unit of electricity.
The total cost is found by multiplying the number of units used by the price per unit.
Charges for this period
Electricity used
Previous 29 125
Latest 29 605
Total 480
Unit charge
10.25p
Total for this period
£49.20

To work out the **cost of electricity** you need to know **how much you've used** (in **units**) and the **price of each unit**. Then it's a simple matter of **multiplying** these two numbers together:

$$\text{Cost} = \text{No. of Units} \times \text{Price per unit}$$

Watch out for the units — you need power in kW (not watts), time in hours (not minutes or seconds) and money in pence or pounds (you can't have both).

Example

How much does it cost to use an **800 W** microwave oven for **6 minutes**? Electricity costs 12.5p per unit.

$E = Pt = 0.8 \times 6/60 = 0.08$ kWh

$\text{Cost} = \text{Units} \times \text{Price} = 0.08 \times 12.5 = 1\text{p}$

Domestic Energy and Fuses

Fuses *Prevent* ***Shocks*** *and* ***Fire***

A **fuse** is a very **fine wire** in a glass tube that's connected between the **live terminal** of the mains supply and an **appliance**. If the **current** in the circuit gets too **big** (bigger than the fuse rating — see below), the fuse wire **melts** and **breaks** the circuit. Fuses should be rated as near as possible but **just higher** than the normal operating current.

The **earth wire** and **fuse** in an appliance with a metal case work together like this:

1) The earth pin in the plug is connected to the **case** of the appliance via the **earth wire**.
2) A **fault** can develop in which the **live** somehow **touches** the case. Then because the case is earthed, a big current flows **in** through the **live**, through the **case** and **out** down the **earth** wire.
3) This **surge** in current blows the fuse, which cuts off the live supply. This prevents electric shocks from the case.

You Can Work Out ***What Fuse*** *to Use from the Appliance's* ***Power Rating***

Most electrical goods have a plate showing their **power rating** and **voltage rating**.
To work out the fuse needed, you have to work out the current that the appliance will normally draw.

Example A toaster is rated at 2.2 kW, 230 V. Find the fuse needed.

You can usually only get fuses with ratings of 3 A, 5 A or 13 A.

Use $P = VI$, and rearrange to give $I = P / V$ = 2200 / 230 = **9.57 A**.
The fuse should be rated just a bit higher than the normal current, so this toaster should have a 13 A fuse.

Practice Questions

Q1 Why aren't joules used on electricity bills? What is used instead?

Q2 What equation links power, energy and time?

Q3 What equation would you use to find the cost of electricity?

Q4 Describe how a fuse works.

Exam Questions

Q1 A vacuum cleaner is rated at 1800 W.

(a) Calculate the energy transferred when the vacuum cleaner is operated for 15 minutes.
Give your answer in: (i) joules, (ii) kilowatt-hours. [2 marks]

(b) Calculate the cost of using the vacuum cleaner for 15 minutes.
Electricity costs 14.6 pence per unit. [1 mark]

Q2 A television is rated at 1500 W, 230 V. Electricity costs 9.8 pence per unit.

(a) Find the fuse needed for the television. [2 marks]

(b) What is the cost of using the television for two and a quarter hours? [2 marks]

When the television is in standby mode, it draws a current of 6.5 mA.

(c) Show that the power of the television in standby mode is approximately 1.5 W. [1 mark]

(d) Estimate the cost of leaving the television on standby overnight (10 hours). [2 marks]

Hurrah — now my toaster won't kill me...

Aren't fuses wonderful things? They're much better than electricity bills, that's for sure. Now, I know there are a couple of equations on these two pages, but you probably met them at GCSE — and they're fairly straightforward, so don't panic. Instead, try the practice exam questions — go on, I know you'll like them — one of them's even about television.

There's resistance everywhere — inside batteries, in all the wires and in the components themselves. No one's for giving current an easy ride.

From now on, I'm assuming that the resistance of the wires in the circuit is zero. In practice, they do have a small resistance.

Batteries have **Resistance**

Resistance comes from **electrons colliding** with **atoms** and **losing energy**.

In a **battery**, **chemical energy** is used to make **electrons move**. As they move, they collide with atoms inside the battery — so batteries **must** have resistance. This is called **internal resistance**.

Internal resistance is what makes **batteries** and **cells warm up** when they're used.

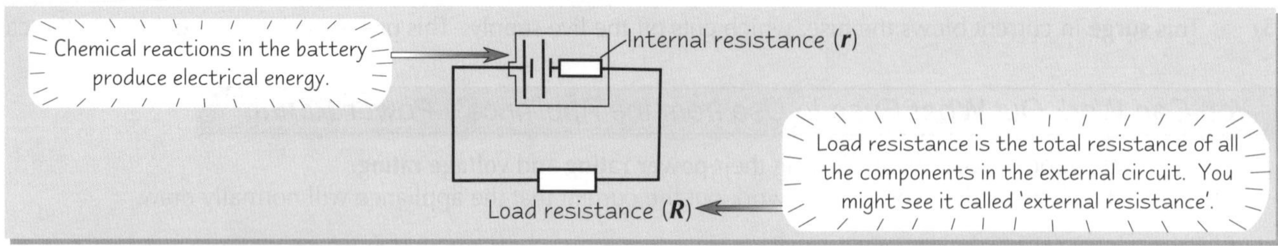

Chemical reactions in the battery produce electrical energy.
Internal resistance (r)
Load resistance is the total resistance of all the components in the external circuit. You might see it called 'external resistance'.
Load resistance (R)

1) The amount of **electrical energy** the battery produces for each **coulomb** of charge is called its **electromotive force** or **e.m.f.** (ε). Be careful — e.m.f. **isn't** actually a force. It's measured in **volts**.
2) The **potential difference** across the **load resistance** (**R**) is the **energy transferred** when **one coulomb** of charge flows through the **load resistance**. This potential difference is called the **terminal p.d.** (V).
3) If there was **no internal resistance**, the **terminal p.d**. would be the **same** as the **e.m.f.** However, in **real** power supplies, there's **always some energy lost** overcoming the internal resistance.
4) The **energy wasted per coulomb** overcoming the internal resistance is called the **lost volts** (v).

Conservation of energy tells us:

energy per coulomb supplied by the source = **energy per coulomb used in load resistance** + **energy per coulomb wasted in internal resistance**

There are Loads of **Calculations with E.m.f.** and **Internal Resistance**

Examiners can ask you to do **calculations** with **e.m.f.** and **internal resistance** in loads of **different** ways. You've got to be ready for whatever they throw at you.

$$\varepsilon = V + v \qquad \varepsilon = I(R + r)$$

$$V = \varepsilon - v \qquad \varepsilon = V + Ir$$

Learn these equations for the exam. Only this one will be on your formula sheet.

These are all basically the **same equation**, just written differently. If you're given enough information you can calculate the e.m.f. (ε), terminal p.d. (V), lost volts (v), current (I), load resistance (R) or internal resistance (r). Which equation you should use depends on what information you've got, and what you need to calculate.

Most Power Supplies Need **Low Internal Resistance**

A **car battery** has to deliver a **really high current** — so it needs to have a **low internal resistance**. The cells used to power a **torch** or a **personal stereo** are the **same**. **Generally**, **batteries** have an **internal resistance** of **less than 1 Ω**.

Since **internal resistance** causes **energy loss**, you'd think **all** power supplies should have a **low internal resistance**.

High voltage power supplies are the **exception**. **HT** (high tension) and **EHT** (extremely high tension) **supplies** are designed with **very high** internal resistances. This means that if they're **accidentally short-circuited** only a **very small current** can flow. Much **safer**.

E.m.f. and Internal Resistance

Use this Circuit to Measure Internal Resistance and E.m.f.

By **changing** the value of ***R*** (**load resistance**) in this circuit and **measuring** the **current** (*I*) and **p.d.** (*V*), you can work out the **internal resistance** of the source.

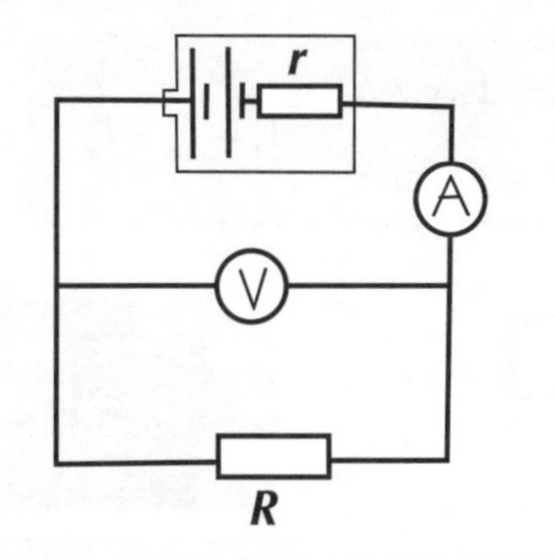

Start with the equation:

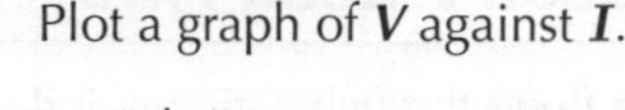

Plot a graph of ***V*** against ***I***.

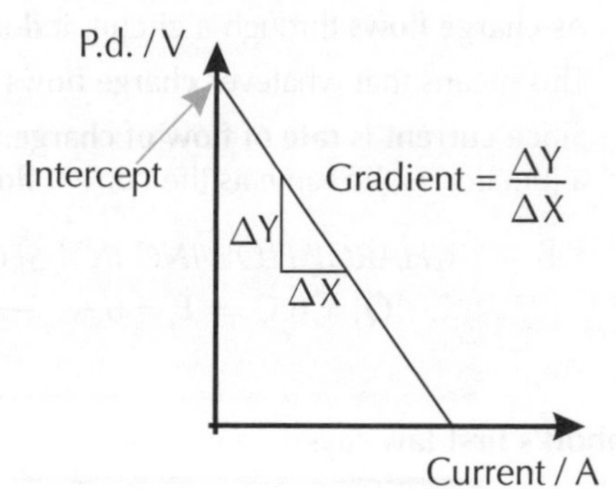

1) Rearrange the equation: $V = -rI + \varepsilon$
2) Since ε and r are constants, that's just the equation of a **straight line** (in the form: $y = mx + c$).
3) So the intercept on the vertical axis is ε.
4) And the gradient is $-r$.

Equation of a straight line
$y = mx + c$
gradient — y-intercept

An **easier** way to **measure** the **e.m.f.** of a **power source** is by connecting a high-resistance **voltmeter** across its **terminals**. A **small current flows** through the **voltmeter**, so there must be some **lost volts** — this means you measure a value **very slightly less** than the **e.m.f.** In **practice** the difference **isn't** usually **significant**.

Practice Questions

Q1 What causes internal resistance?

Q2 What is meant by 'lost volts'?

Q3 What is the difference between e.m.f. and terminal p.d.?

Q4 Write the equation used to calculate the terminal p.d. of a power supply.

Exam Questions

Q1 A large battery with an internal resistance of 0.8 Ω and e.m.f. 24 V is used to power a dentist's drill with resistance 4 Ω.

(a) Calculate the current in the circuit when the drill is connected to the power supply. [2 marks]

(b) Calculate the voltage across the drill while it is being used. [1 mark]

Q2 A student mistakenly connects a 10 Ω ray box to an HT power supply of 500 V. The ray box does not light, and the student measures the current flowing to be only 50 mA.

(a) Calculate the internal resistance of the HT power supply. [2 marks]

(b) Explain why this is a sensible internal resistance for an HT power supply. [2 marks]

You're UNBELIEVABLE... [Frantic air guitar]... Ueuuurrrrghhh... Yeah...

Wanting power supplies to have a low internal resistance makes sense — you wouldn't want your MP3 player battery melting if you listened to music for more than half an hour. Make sure you know your e.m.f. equations — they're an exam fave. A good way to get them learnt is to keep trying to get from one equation to another... dull, but it can help.

Conservation of Energy & Charge in Circuits

There are some things in Physics that are so fundamental that you just have to accept them. Like the fact that there's loads of Maths in it. And that energy is conserved. And that Physicists get more homework than everyone else.

Charge Doesn't 'Leak Away' Anywhere — it's Conserved

1) As **charge flows** through a circuit, it **doesn't** get **used up** or **lost**.
2) This means that whatever **charge flows into** a junction will **flow out** again.
3) Since **current** is **rate of flow of charge**, it follows that whatever **current flows into** a junction is the same as the current **flowing out** of it.

e.g. *CHARGE FLOWING IN 1 SECOND*

$Q_1 = 6\text{ C} \Rightarrow I_1 = 6\text{ A}$ → $Q_2 = 2\text{ C} \Rightarrow I_2 = 2\text{ A}$ and $Q_3 = 4\text{ C} \Rightarrow I_3 = 4\text{ A}$

$I_1 = I_2 + I_3$

Kirchhoff's first law says:

The total **current entering a junction** = the total **current leaving it.**

Energy conservation is vital.

Energy is Conserved too

1) **Energy is conserved**. You already know that. In **electrical circuits**, **energy** is **transferred round** the circuit. Energy **transferred to** a charge is **e.m.f.**, and energy **transferred from** a charge is **potential difference**.
2) In a **closed loop**, these two quantities must be **equal** if energy is conserved (which it is).

Kirchhoff's second law says:

The **total e.m.f.** around a **series circuit** = the **sum** of the **p.d.s** across each component. (or $\varepsilon = \Sigma IR$ in symbols)

Exam Questions get you to Apply Kirchhoff's Laws to Combinations of Resistors

A **typical exam question** will give you a **circuit** with bits of information missing, leaving you to fill in the gaps. Not the most fun... but on the plus side you get to ignore any internal resistance stuff (unless the question tells you otherwise)... hurrah. You need to remember the **following rules**:

SERIES Circuits

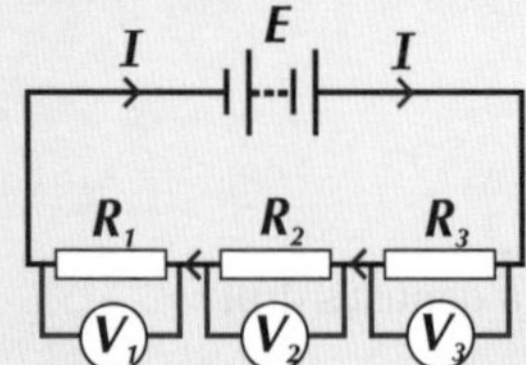

1) **same current** at **all points** of the circuit (since there are no junctions)
2) **e.m.f. split** between **components** (by Kirchhoff's 2nd law), so: $E = V_1 + V_2 + V_3$
3) $V = IR$, so if I is constant: $IR_{total} = IR_1 + IR_2 + IR_3$
4) cancelling the Is gives:

$$R_{total} = R_1 + R_2 + R_3$$

PARALLEL Circuits

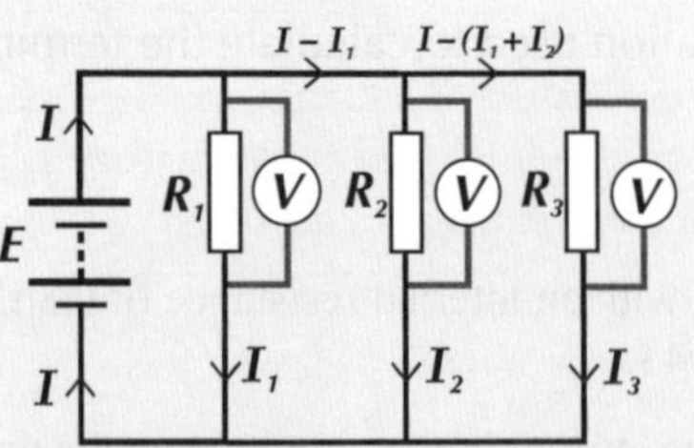

1) **current** is **split** at each **junction**, so: $I = I_1 + I_2 + I_3$
2) **same p.d.** across **all components** (three separate loops — within each loop the e.m.f. equals sum of individual p.d.s)
3) so, $V/R_{total} = V/R_1 + V/R_2 + V/R_3$
4) cancelling the Vs gives:

$$1/R_{total} = 1/R_1 + 1/R_2 + 1/R_3$$

...and there's an example on the next page to make sure you know what to do with all that...

Conservation of Energy & Charge in Circuits

Worked Exam Question

A battery of e.m.f. 16 V and negligible internal resistance is connected in a circuit as shown:

a) Show that the group of resistors between X and Y could be replaced by a single resistor of resistance 15 Ω.

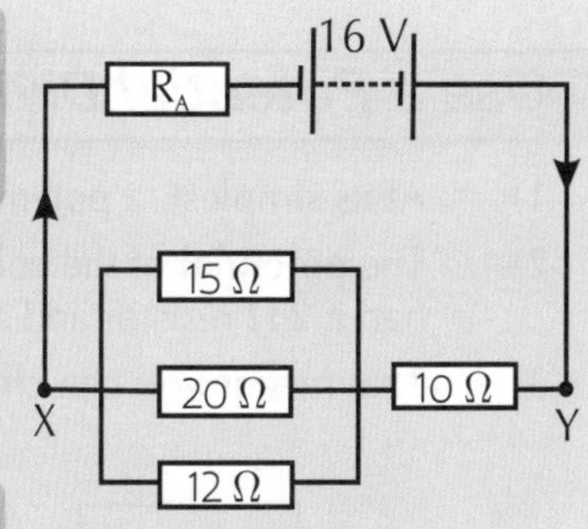

You can find the **combined resistance** of the 15 Ω, 20 Ω and 12 Ω resistors using:

$1/R = 1/R_1 + 1/R_2 + 1/R_3 = 1/15 + 1/20 + 1/12 = 1/5 \quad \Rightarrow R = 5\ \Omega$

So **overall resistance** between **X** and **Y** can be found by $R = R_1 + R_2 = 5 + 10 = \mathbf{15\ \Omega}$

b) If $R_A = 20\ \Omega$:
(i) calculate the potential difference across R_A,

Careful — there are a few steps here. You need the p.d. across R_A, but you don't know the current through it. So start there:
total resistance in circuit = 20 + 15 = 35 Ω, **so** current through R_A can be found using $I = V_{total}/R_{total}$:

$I = 16/35$ A

then you can use $V = IR_A$ to find the p.d. across R_A: $V = 16/35 \times 20 = \mathbf{9.1\ V}$

(ii) calculate the current in the 15 Ω resistor.

You know the **current flowing** into the group of three resistors and out of it, but not through the individual branches. But you know that their **combined resistance** is **5** Ω (from part a) so you can work out the p.d. across the group:

$V = IR = 16/35 \times 5 = 16/7$ V

The p.d. across the **whole group** is the same as the p.d. across each **individual resistor**, so you can use this to find the current through the 15 Ω resistor:

$I = V/R = (16/7) / 15 = \mathbf{0.15\ A}$

Practice Questions

Q1 State Kirchhoff's laws.

Q2 Find the current through and potential difference across each of two 5 Ω resistors when they are placed in a circuit containing a 5 V battery, and are wired: a) in series, b) in parallel.

Exam Question

Q1 For the circuit on the right:

(a) Calculate the total effective resistance of the three resistors in this combination. [2 marks]

(b) Calculate the main current, I_3. [2 marks]

(c) Calculate the potential difference across the 4 Ω resistor. [1 mark]

(d) Calculate the potential difference across the parallel pair of resistors. [1 mark]

(e) Using your answer from 1 (d), calculate the currents I_1 and I_2. [2 marks]

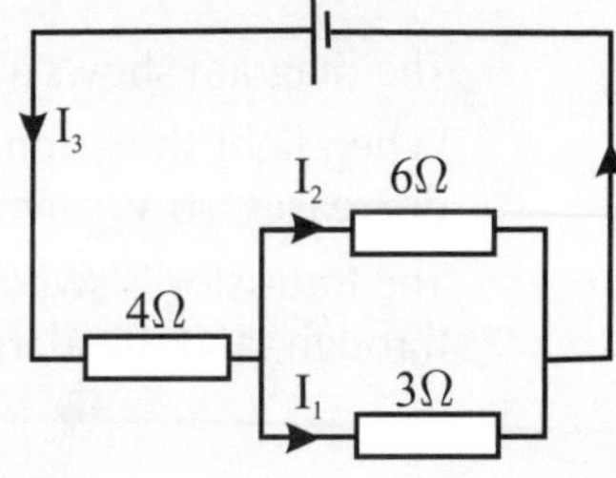

This is a very purple page — needs a bit of yellow I think...

V = IR is the formula you'll use most often in these questions. Make sure you know whether you're using it on the overall circuit, or just one specific component. It's amazingly easy to get muddled up — you've been warned.

The Potential Divider

I remember the days when potential dividers were pretty much the hardest thing they could throw at you. Then along came AS Physics. Hey ho.

Anyway, in context this doesn't seem too hard now, so get stuck in.

Use a Potential Divider to get a Fraction of a Source Voltage

1) At its simplest, a **potential divider** is a circuit with a **voltage source** and a couple of **resistors** in series.
2) The **potential** of the voltage source (e.g. a power supply) is **divided** in the **ratio** of the **resistances**. So, if you had a **2 Ω** resistor and a **3 Ω** resistor, you'd get **2/5** of the p.d. across the **2 Ω** resistor and **3/5** across the **3 Ω**.
3) That means you can **choose** the **resistances** to get the **voltage** you **want** across one of them.

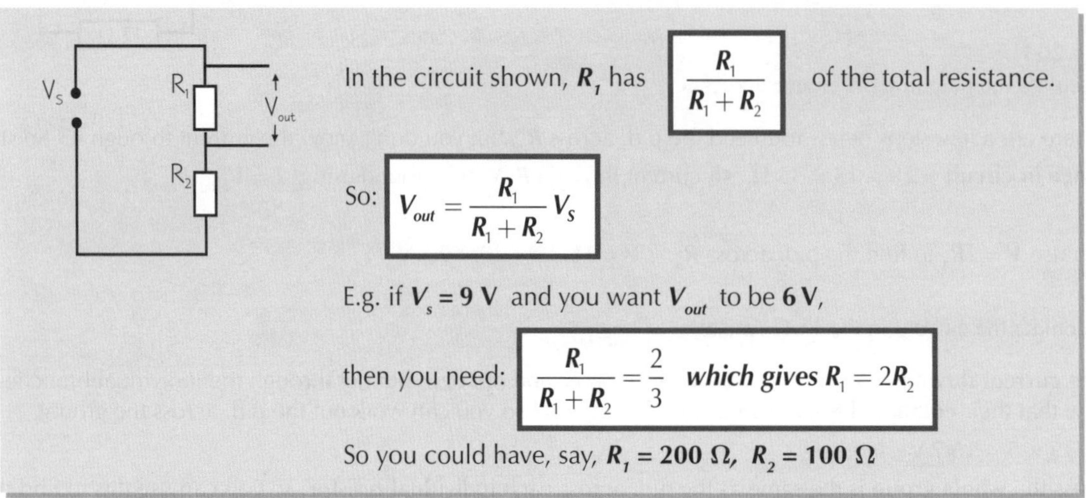

In the circuit shown, R_1 has $\frac{R_1}{R_1+R_2}$ of the total resistance.

So: $$V_{out} = \frac{R_1}{R_1+R_2} V_s$$

E.g. if V_s = **9 V** and you want V_{out} to be **6 V**,

then you need: $$\frac{R_1}{R_1+R_2} = \frac{2}{3} \quad \textit{which gives } R_1 = 2R_2$$

So you could have, say, R_1 **= 200 Ω**, R_2 **= 100 Ω**

4) This circuit is mainly used for **calibrating voltmeters**, which have a **very high resistance**.
5) If you put something with a **relatively low resistance** across R_1 though, you start to run into **problems**. You've **effectively** got **two resistors** in **parallel**, which will **always** have a **total** resistance **less** than R_1. That means that V_{out} will be **less** than you've calculated, and will depend on what's connected across R_1. Hrrumph.

Add an LDR or Thermistor for a Light or Temperature Switch

1) A **light-dependent resistor** (LDR) has a very **high resistance** in the **dark**, but a **lower resistance** in the **light**.
2) An **NTC thermistor** has a **high resistance** at **low temperatures**, but a much **lower resistance** at **high temperatures** (it varies in the opposite way to a normal resistor, only much more so).
3) Either of these can be used as one of the **resistors** in a **potential divider**, giving an **output voltage** that **varies** with the **light level** or **temperature**.
4) Add a **transistor** and you've got yourself a **switch**, e.g. to turn on a light or a heating system.

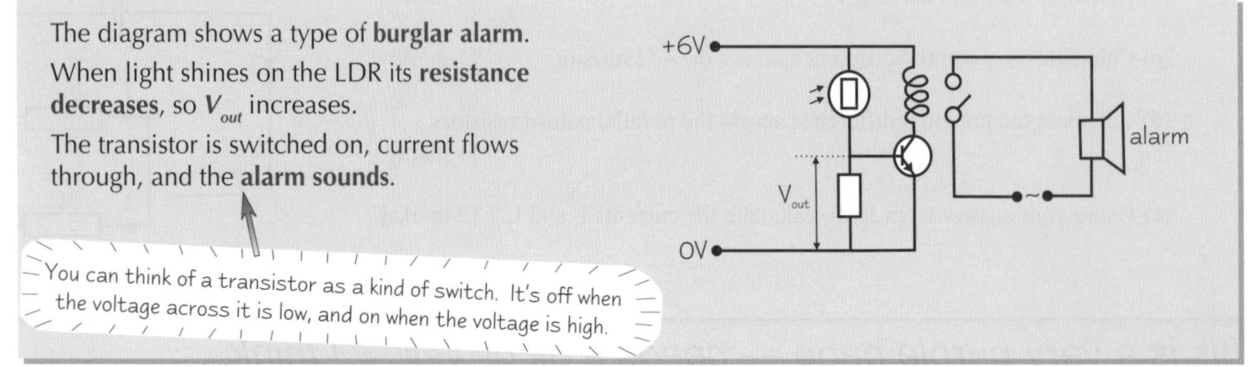

The diagram shows a type of **burglar alarm**. When light shines on the LDR its **resistance decreases**, so V_{out} increases. The transistor is switched on, current flows through, and the **alarm sounds.**

You can think of a transistor as a kind of switch. It's off when the voltage across it is low, and on when the voltage is high.

The Potential Divider

A ***Potentiometer*** *uses a* ***Variable Resistor*** *to give a* ***Variable Voltage***

1) A **potentiometer** has a variable resistor replacing R_1 and R_2 of the potential divider, but it uses the **same idea** (it's even sometimes **called** a potential divider just to confuse things).
2) You move a **slider** or turn a knob to **adjust** the **relative sizes** of R_1 and R_2. That way you can vary V_{out} from **0 V** up to the source voltage.
3) This is dead handy when you want to be able to **change** a **voltage continuously**, like in the **volume control** of a stereo.

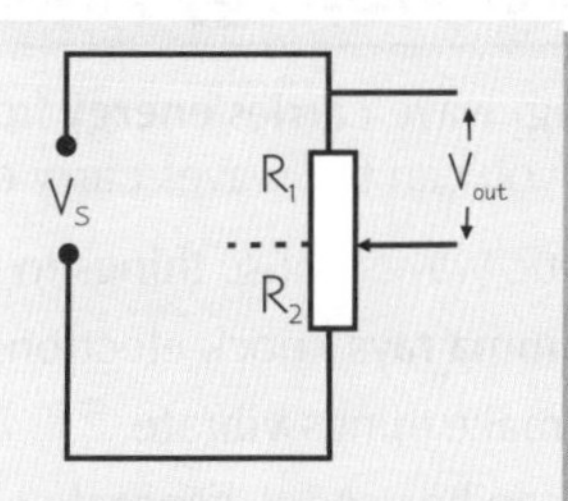

Here, V_S is replaced by the input signal (e.g. from a CD player) and V_{out} is the output to the amplifier and loudspeaker.

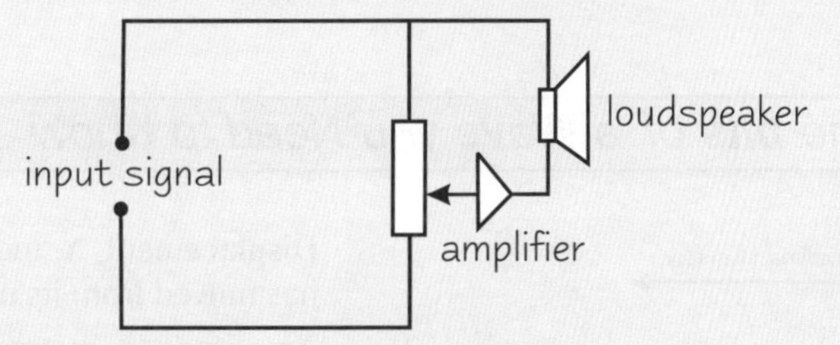

Practice Questions

Q1 Look at the burglar alarm circuit on page 56. How could you change the circuit so that the alarm sounds when the light level decreases?

Q2 The LDR in the burglar alarm circuit has a resistance of 300 Ω when light and 900 Ω when dark. The fixed resistor has a value of 100 Ω. Show that V_{out} (light) = 1.5 V and V_{out} (dark) = 0.6 V.

Exam Questions

Q1 In the circuit on the right, all the resistors have the same value. Calculate the p.d. between:

(i) A and B. [1 mark]

(ii) A and C. [1 mark]

(iii) B and C. [1 mark]

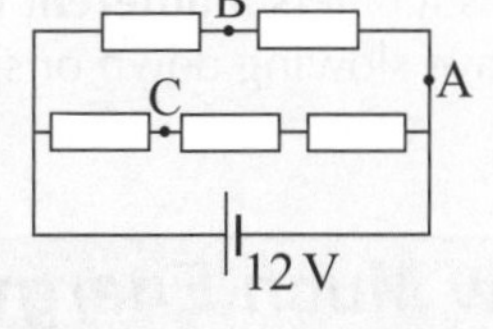

Q2 Look at the circuit on the right.

(a) Calculate the p.d. between A and B as shown by a high resistance voltmeter placed between the two points. [1 mark]

(b) A 40 Ω resistor is now placed between points A and B. Calculate the p.d. across AB and the current flowing through the 40 Ω resistor. [4 marks]

30 Ω
10 Ω
12 V
A
50 Ω
B

OI... YOU... [bang bang bang]... turn that potentiometer down...

You'll probably have to use a potentiometer in every experiment you do with electricity from now on in, so you'd better get used to them. I can't stand the things myself, but then lab and me don't mix — far too technical.

The Nature of Waves

Aaaah... playing with slinky springs and waggling ropes about. It's all good clean fun as my mate Richard used to say...

A **Wave Transfers Energy** Away from Its Source

A **progressive** (moving) wave carries **energy** from one place to another **without transferring any material**. Here are some ways you can tell waves carry energy:

1) Electromagnetic waves cause things to **heat up**.
2) **X-rays** and **gamma rays** knock electrons out of their orbits, causing **ionisation**.
3) Loud **sounds** make things **vibrate**.
4) **Wave power** can be used to **generate electricity**.
5) Since waves carry energy away, the **source** of the wave **loses energy**.

Here are all the **bits** of a **Wave** you Need to Know

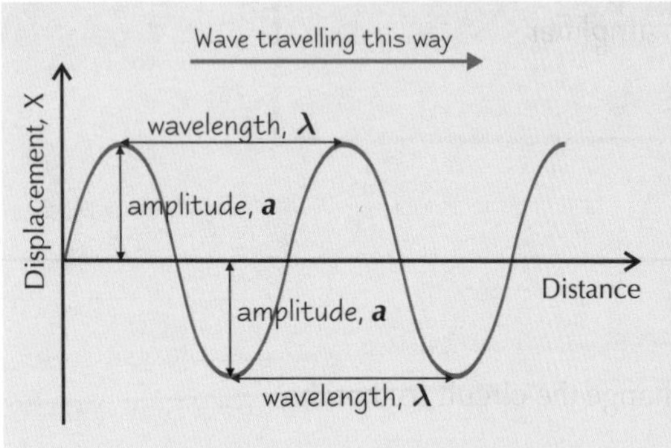

1) **Displacement, *X*,** metres — how far a **point** on the wave has **moved** from its **undisturbed position.**
2) **Amplitude, a,** metres — **maximum displacement.**
3) **Wavelength, λ,** metres — the **length** of **one whole wave,** from **crest** to **crest** or **trough** to **trough.**
4) **Period, *T*,** seconds — the **time taken** for a **whole vibration.**
5) **Frequency, *f*,** hertz — the **number** of **vibrations per second** passing a given **point.**
6) **Phase difference** — the amount by which **one wave lags behind another** wave. **Measured** in **degrees** or **radians.** See page 64.

Waves Can Be **Reflected** and **Refracted**

Reflection — the wave is **bounced back** when it **hits a boundary**. E.g. you can see the reflection of light in mirrors. The reflection of water waves can be demonstrated in a ripple tank.

Refraction — the wave **changes direction** as it enters a **different medium**.
The change in direction is a result of the wave slowing down or speeding up.

Intensity is a Measure of **How Much Energy** a Wave is Carrying

1) When you talk about "**brightness**" for light or "**loudness**" for sound, what you really mean is **how much light** or **sound** energy hits your eyes or your ears **per second**.
2) The scientific measure of this is **intensity**.

> Intensity is the **rate of flow** of **energy** per **unit area** at **right angles** to the **direction of travel** of the wave. It's measured in **Wm^{-2}**.

Intensity is **Proportional** to the **Square** of the **Amplitude** of the **Wave**

1) This comes from the fact that **intensity** is **proportional** to **energy**, and the energy of a wave depends on the square of the **amplitude**.
2) From this you can tell that for a **vibrating source** it takes four times as much energy to double the size of the vibrations.

The Nature of Waves

The *Frequency* is the *Inverse* of the *Period*

$$\textit{Frequency} = \frac{1}{\textit{period}}$$

It's that simple.
Get the **units** straight: **1 Hz = 1 s^{-1}**.

Wave Speed, Frequency and *Wavelength* are Linked by the *Wave Equation*

Wave speed can be measured just like the speed of anything else:

$$\text{Speed } (v) = \frac{\text{distance moved } (d)}{\text{time taken } (t)}$$

Remember, you're not measuring how fast a physical point (like one molecule of rope) moves. You're measuring how fast a point on the **wave pattern** moves.

Learn the **Wave Equation**...

Speed of wave (v) = wavelength (λ) × frequency (f)

$$v = \lambda f$$

You need to be able to rearrange this equation for v, λ or f.

... and How to **Derive** it

You can work out the **wave equation** by imagining **how long** it takes for the **crest** of a wave to **move** across a **distance** of **one wavelength**. The **distance travelled** is **λ**. **By definition**, the **time taken** to travel **one whole wavelength** is the **period** of the wave, which is equal to **1/*f***.

$$\text{Speed } (v) = \frac{\text{distance moved } (d)}{\text{time taken } (t)} \Longrightarrow \text{Speed } (v) = \frac{\text{distance moved } (\lambda)}{\text{time taken } (1/f)}$$

Learn to recognise when to use $v = \lambda f$ and when to use $v = d/t$. Look at which variables are mentioned in the question.

Practice Questions

Q1 Does a wave carry matter **or** energy from one place to another?

Q2 Diffraction and interference are two wave properties. Write down two more.

Q3 Write down the relationship between the amplitude of a wave and its intensity.

Q4 Give the units of frequency, displacement and amplitude.

Q5 Write down the equation connecting v, λ and f.

Exam Question

Q1 A buoy floating on the sea takes 6 seconds to rise and fall once (complete a full period of oscillation). The difference in height between the buoy at its lowest and highest points is 1.2 m, and waves pass it at a speed of 3 ms^{-1}.

(a) How long are the waves? [2 marks]

(b) What is the amplitude of the waves? [1 mark]

Learn the wave equation and its derivation — pure poetry...

This isn't too difficult to start you off — most of it you'll have done at GCSE anyway. But once again, it's a whole bunch of equations to learn, and you won't get far without learning them. Yada yada.

There are different types of wave — and the difference is easiest to see using a slinky. Try it — you'll have hours of fun.

*In **Transverse Waves** the **Vibration** is at **Right Angles** to the **Direction** of Travel*

All **electromagnetic waves** are **transverse**. Other examples of transverse waves are **ripples** on water and waves on **ropes**.

There are **two** main ways of **drawing** transverse waves:

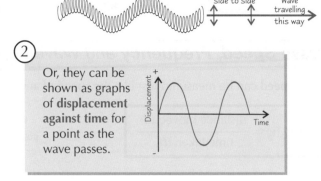

Vibrations from side to side
Wave travelling this way

1) They can be shown as **graphs** of **displacement** against **distance along the path of the wave**.

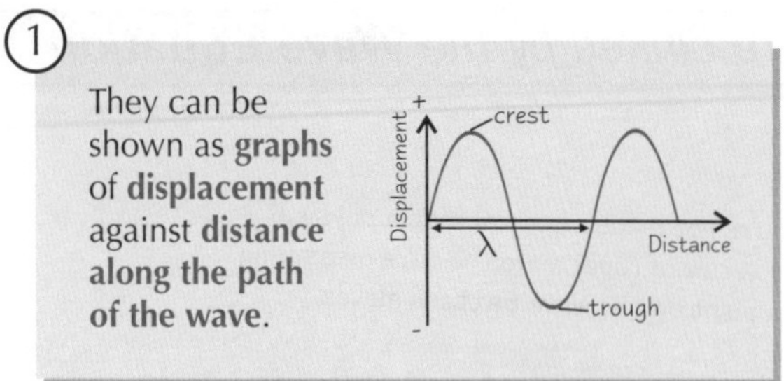

2) Or, they can be shown as graphs of **displacement against time** for a point as the wave passes.

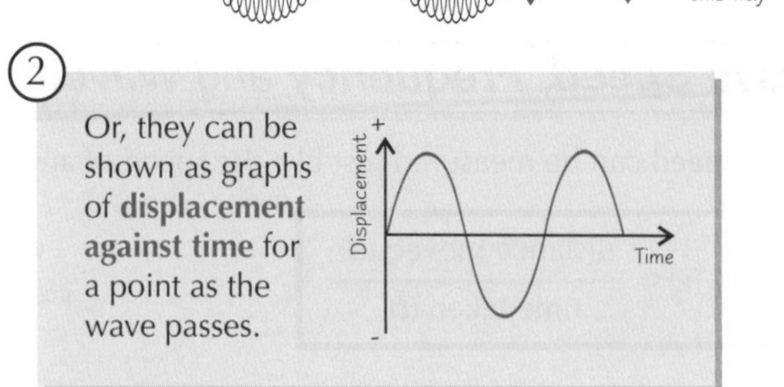

Both sorts of graph often give the **same shape**, so make sure you check out the label on the **x-axis**. Displacements **upwards** from the centre line are given a **+ sign**. Displacements downwards are given a **– sign**.

*In **Longitudinal Waves** the **Vibrations** are **Along** the Direction of Travel*

The most **common** example of a **longitudinal wave** is **sound**. A sound wave consists of alternate **compressions** and **rarefactions** of the **medium** it's travelling through. (That's why sound can't go through a vacuum.) Some types of **earthquake shock waves** are also longitudinal.

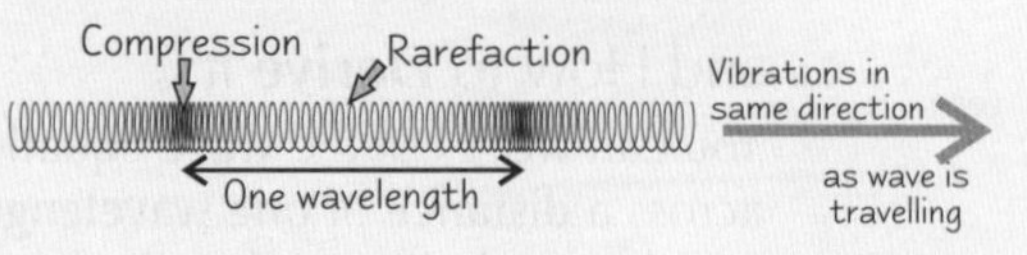

It's hard to **represent** longitudinal waves **graphically**. You'll usually see them plotted as **displacement** against **time**. These can be **confusing** though, because they look like a **transverse wave**.

*A **Polarised Wave** only **Oscillates** In One Direction*

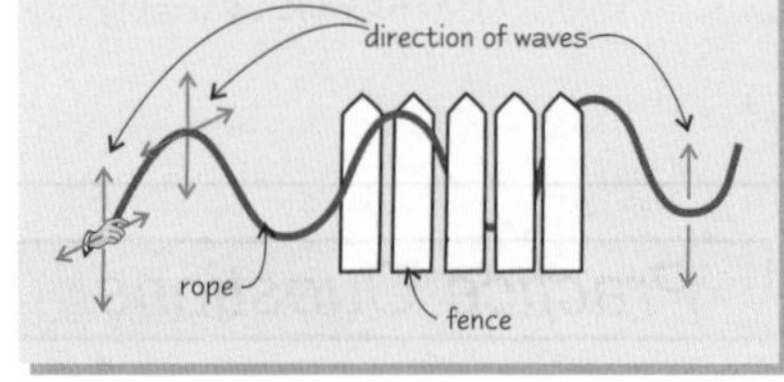

1) If you **shake a rope** to make a **wave** you can move your hand **up and down** or **side to side** or in a **mixture** of directions — it still makes a **transverse wave**.
2) But if you try to pass **waves in a rope** through a **vertical fence**, the wave will only get through if the **vibrations** are **vertical**. The fence filters out vibration in other directions. This is called **polarising** the wave.
3) Ordinary **light waves** are a mixture of **different directions** of **vibration**. (The things vibrating are electric and magnetic fields.) A **polarising filter** only transmits vibrations in one direction.
4) If you have two polarising filters at **right angles** to each other, then **no** light will get through.
5) Polarisation **can only happen** for **transverse** waves. The fact that you can polarise light is one **proof** that it's a transverse wave.

*When **Light Reflects** it is **Partially Polarised***

1) Rotating a **polarising filter** in a beam of light shows the fraction of the light that is vibrating in each **direction**.
2) If you direct a beam of unpolarised light at a reflective surface then view the **reflected ray** through a polarising filter, the intensity of light leaving the filter **changes** with the **orientation** of the filter.
3) The intensity changes because light is **partially polarised** when it is **reflected**.
4) This effect is used to remove **unwanted reflections** in photography and in **Polaroid sunglasses** to remove **glare**.

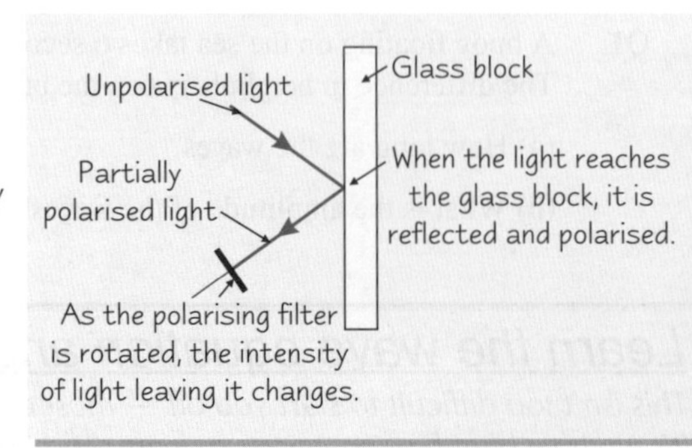

Unpolarised light
Glass block
Partially polarised light
When the light reaches the glass block, it is reflected and polarised.
As the polarising filter is rotated, the intensity of light leaving it changes.

Longitudinal and Transverse Waves

Materials Can **Rotate** the **Plane of Polarisation**

The **plane** in which a wave moves and **vibrates** is called the **plane of polarisation** — e.g. the rope on the last page was polarised in the **vertical** plane by the fence. Some **materials** (e.g. crystals) **rotate** the plane of polarised light. You can **measure** how much a material rotates the plane of polarised light using two **polarising filters**:

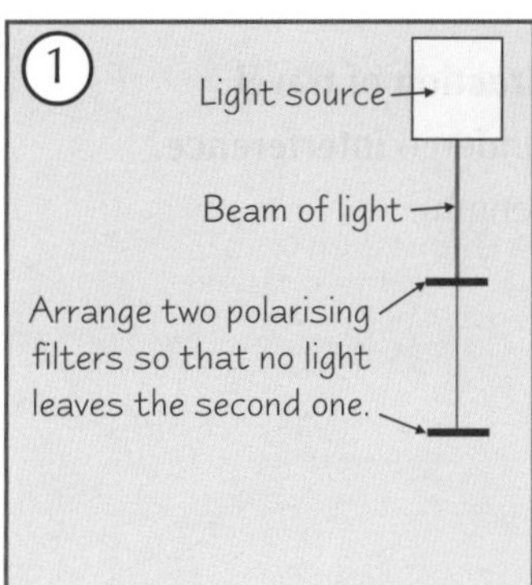

1
Light source
Beam of light
Arrange two polarising filters so that no light leaves the second one.

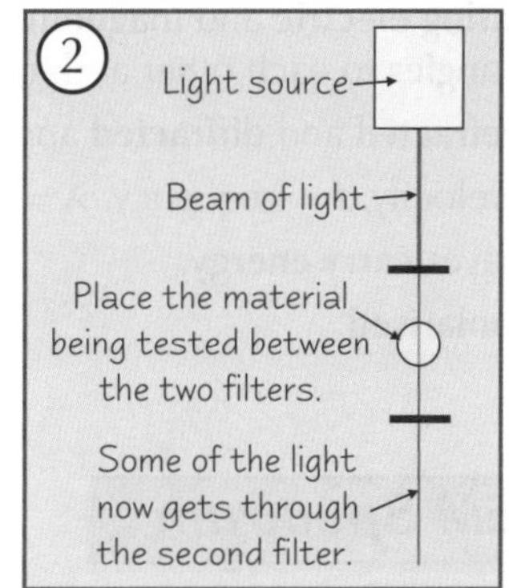

2
Light source
Beam of light
Place the material being tested between the two filters.
Some of the light now gets through the second filter.

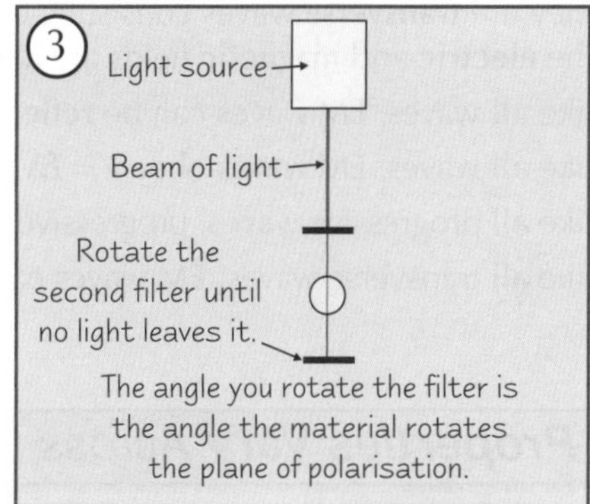

3
Light source
Beam of light
Rotate the second filter until no light leaves it.
The angle you rotate the filter is the angle the material rotates the plane of polarisation.

Rotating the **Plane of Polarisation** Affects the **Intensity**

You've seen that passing light through a polarising filter **rotates** its **plane of polarisation**, but it also changes the **amplitude** and **intensity** of the transmitted wave.

The **amplitude** of the **transmitted** wave is the **component** of the **incident** wave in the direction of the **new plane** of polarisation.

$$A = A_0 \cos\theta$$

Where A is the amplitude of the **transmitted** wave, A_0 is the amplitude of the **incident** wave and θ (theta) is the **angle** the plane has been rotated.

The **intensity** of the **transmitted** light is **proportional** to the **amplitude squared** — this is **Malus' law**.

$$I = I_0 \cos^2\theta$$

Where I is the intensity of transmitted light, I_0 is the intensity of incident light and θ (theta) is the **angle** the plane has been rotated.

Practice Questions

Q1 Give examples of a transverse wave and a longitudinal wave.

Q2 What is a polarised wave? How can you polarise a wave?

Q3 What is Malus' law and what is it used for?

Exam Questions

Q1 In an experiment, light is shone through a disc of a crystal called "Iceland spar". The beam of light is less bright when it emerges from the crystal than when it enters. Next, a second identical disc of Iceland spar is placed in front of the first. The first disc is held steady while the second is rotated (in the plane of the disc). The intensity of light emerging changes as the second disc rotates. At two points in each rotation, no light gets through at all.

(a) Explain the results of these experiments. You may use a diagram to help your answer. [5 marks]

(b) When the second disc is rotated to angle α, the intensity of light emerging from the second disc is half the value it was when it left the first disc. Calculate angle α. [3 marks]

Q2 Give one example of an application of polarisation and explain how it works. [2 marks]

Caution — rotating the plane may cause nausea...

The waves broadcast from TV or radio transmitters are polarised. So you have to line up the receiving aerial with the transmitting aerial to receive the signal properly. It's one reason why the TV picture's lousy if the aerial gets knocked.

There's nothing really deep and meaningful to understand on this page — just a load of facts to learn I'm afraid.

All **Electromagnetic Waves** *Have Some* **Properties** *In Common*

1) They travel in a **vacuum** at a **speed** of **2.998 × 10^8 ms^{-1}**, and at slower speeds in other media.
2) They are **transverse** waves consisting of **vibrating electric** and **magnetic fields**. The **electric** and **magnetic** fields are at **right angles** to each other and to the **direction of travel**.
3) Like all waves, EM waves can be **reflected**, **refracted** and **diffracted** and can undergo **interference**.
4) Like all waves, EM waves obey $v = f\lambda$ (v = velocity, f = frequency, λ = wavelength).
5) Like all progressive waves, progressive EM waves **carry energy**.
6) Like all transverse waves, EM waves can be **polarised**.

Some **Properties Vary** *Across the* **EM Spectrum**

EM waves with different wavelengths behave differently in some respects. The spectrum is split into seven categories: **radio waves**, **microwaves**, **infrared**, **visible light**, **ultraviolet**, **X-rays** and **gamma rays**.

1) The longer the wavelength, the more **obvious** the wave characteristics — e.g., long radio waves diffract round hills.
2) **Energy** is directly proportional to **frequency**. **Gamma rays** have the **highest energy**; **radio waves** the **lowest**.
3) The **higher** the **energy**, in general the more **dangerous** the wave.
4) The **lower the energy** of an EM wave, the **further from the nucleus** it comes from. **Gamma radiation** comes from inside the **nucleus**. **X-rays to visible light** come from energy-level transitions in **atoms** (see p. 78). **Infrared** radiation and **microwaves** are associated with **molecules**. **Radio waves** come from oscillations in **electric fields**.

The **Properties** of an **EM Wave** Change with **Wavelength**

Type	Approximate wavelength / m	Penetration	Uses
Radio waves	10^{-1} — 10^{6}	Pass through matter.	Radio transmissions.
Microwaves	10^{-3} — 10^{-1}	Mostly pass through matter, but cause some heating.	Radar. Microwave cookery. TV transmissions.
Infrared (IR)	7×10^{-7} — 10^{-3}	Mostly absorbed by matter, causing it to heat up.	Heat detectors. Night-vision cameras. Remote controls. Optical fibres.
Visible light	4×10^{-7} — 7×10^{-7}	Absorbed by matter, causing some heating effect.	Human sight. Optical fibres.
Ultraviolet (UV)	10^{-8} — 4×10^{-7}	Absorbed by matter. Slight ionisation.	Sunbeds. Security markings that show up in UV light.
X-rays	10^{-13} — 10^{-8}	Mostly pass through matter, but cause ionisation as they pass.	To see damage to bones and teeth. Airport security scanners. To kill cancer cells.
Gamma rays	10^{-16} — 10^{-10}	Mostly pass through matter, but cause ionisation as they pass.	Irradiation of food. Sterilisation of medical instruments. To kill cancer cells.

The Electromagnetic Spectrum

Different Types of EM Wave Have Different **Effects** on the **Body**

Type	Production	Effect on human body
Radio waves	Oscillating electrons in an aerial	No effect.
Microwaves	Electron tube oscillators. Masers.	Absorbed by water — danger of cooking human body*.
Infrared (IR)	Natural and artificial heat sources.	Heating. Excess heat can harm the body's systems.
Visible light	Natural and artificial light sources.	Used for sight. Too bright a light can damage eyes.
Ultraviolet (UV)	e.g. the Sun.	Tans the skin. Can cause skin cancer and eye damage.
X-rays	Bombarding metal with electrons.	Cancer due to cell damage. Eye damage.
Gamma rays	Radioactive decay of the nucleus.	Cancer due to cell damage. Eye damage.

1) UV radiation is split into categories based on frequency — **UV-A**, **UV-B** and **UV-C**.
2) **UV-A** has the **lowest** frequency and is the **least damaging**, although it's thought to be a significant cause of **skin aging**.
3) Higher-frequency **UV-B** is more dangerous. It can be **absorbed** by DNA molecules, causing **mutations** which can lead to **cancer**. UV-B is responsible for **sunburn** too.
4) **UV-C** has a high enough frequency to be **ionising** — it carries enough energy to knock electrons off atoms. This can cause **cell mutation** or **destruction**, and **cancer**. It's almost **entirely blocked** by the ozone layer, though.
5) **Dark** skin gives some protection from UV rays, stopping them reaching more vulnerable tissues below. So **tanning** is a protection mechanism — **UV-A** triggers the release of melanin (a brown pigment) in the skin.
6) **Sunscreens** provide some protection from UV in sunlight. The **Sun Protection Factor** (**SPF**) of the sunscreen tells you how well it protects against **UV-B** radiation. It **doesn't** tell you anything about the UV-A protection though. Many modern sunscreens include tiny particles of **zinc oxide** and **titanium dioxide** to block UV-A.

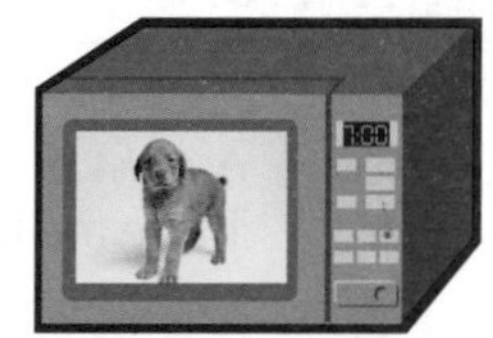

* Or small animals.

Practice Questions

Q1 What are the main practical uses of infrared radiation?

Q2 Which types of electromagnetic radiation have the highest and lowest energies?

Q3 What is the significance of the speed 2.998×10^8 ms^{-1}?

Q4 Why are microwaves dangerous?

Q5 How does the energy of an EM wave vary with frequency?

Exam Questions

Q1 In a vacuum, do X-rays travel faster, slower or at the same speed as visible light? Explain your answer. [2 marks]

Q2 (a) Describe briefly the physics behind a practical use of X rays. [2 marks]

(b) What is the difference between gamma rays and X-rays? [2 marks]

Q3 Give an example of a type of electromagnetic wave causing a hazard to health. [2 marks]

I've got UV hair...

No really I have. It's great. It's purple. And it's got shiny glittery white bits in it. Aaaanyway... moving swiftly on. Loads of facts to learn on these pages. You probably know most of this from GCSE anyway, but make sure you know it well enough to answer a question on it in the exam. Not much fun, but... there you go.

When two waves get together, it can be either really impressive or really disappointing.

Superposition Happens When **Two** or **More** Waves **Pass Through** Each Other

1) At the **instant** the waves **cross**, the **displacements** due to each wave **combine**. Then **each wave** goes on its merry way. You can **see** this if **two pulses** are sent **simultaneously** from each end of a rope.
2) The **principle of superposition** says that when two or more **waves cross**, the **resultant** displacement equals the **vector sum** of the **individual** displacements.

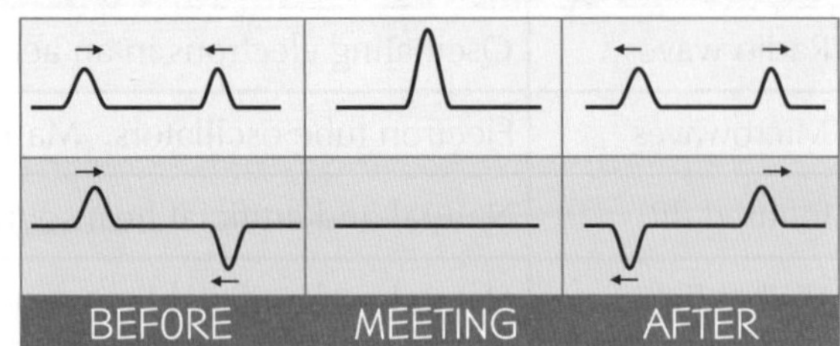

"Superposition" means "one thing on top of another thing". You can use the same idea in reverse — a complex wave can be separated out mathematically into several simple sine waves of various sizes.

Interference can be **Constructive** or **Destructive**

1) A **crest** plus a **crest** gives a **big crest**. A **trough** plus a **trough** gives a **big trough**. These are both examples of **constructive interference**.
2) A **crest** plus a **trough** of **equal size** gives... **nothing**. The two displacements **cancel each other out** completely. This is called **destructive interference**.
3) If the **crest** and the **trough** aren't the **same size**, then the destructive interference **isn't total**. For the interference to be **noticeable**, the two **amplitudes** should be **nearly equal**.

Graphically, you can superimpose waves by adding the individual displacements at each point along the x-axis, and then plotting them.

In **Phase** *Means In* **Step** — *Two Points* **In Phase** *Interfere* **Constructively**

1) Two points on a wave are **in phase** if they are both at the **same point** in the **wave cycle**. Points in phase have the **same displacement** and **velocity**.
   On the graph, points **A** and **B** are **in phase**; points **A** and **C** are **out of phase**.

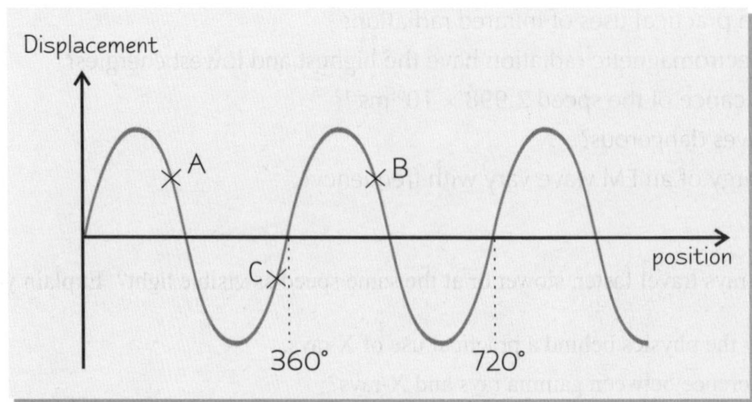

2) It's mathematically **handy** to show one **complete cycle** of a wave as an **angle of 360° (2π radians)**.
   Two points with a **phase difference** of **zero** or a **multiple of 360°** are **in phase**.
   Points with a **phase difference** of **odd-number multiples** of **180° (π radians)** are **exactly out of phase**.
3) You can also talk about two **different waves** being **in phase**. **In practice** this happens because **both** waves came from the **same oscillator**. In **other** situations there will nearly always be a **phase difference** between two waves.

Superposition and Coherence

To Get **Interference Patterns** the **Two Sources** Must Be **Coherent**

Interference **still happens** when you're observing waves of **different wavelength** and **frequency** — but it happens in a **jumble**. In order to get clear **interference patterns**, the two or more sources must be **coherent**.

Two sources are **coherent** if they have the **same wavelength** and **frequency** and a **fixed phase difference** between them.

In exam questions at AS, the 'fixed phase difference' is almost certainly going to be zero. The two sources will be in phase.

Constructive or **Destructive** Interference Depends on the **Path Difference**

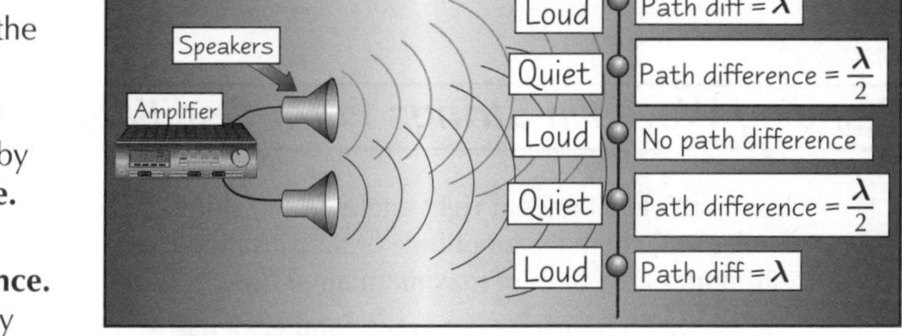

1) Whether you get **constructive** or **destructive** interference at a **point** depends on how **much further one wave** has travelled than the **other wave** to get to that point.
2) The **amount** by which the path travelled by one wave is **longer** than the path travelled by the other wave is called the **path difference.**
3) At **any point an equal distance** from both sources you will get **constructive interference.** You also get constructive interference at any point where the **path difference** is a **whole number of wavelengths**. At these points the two waves are **in phase** and **reinforce** each other. But at points where the path difference is **half a wavelength**, **one and a half** wavelengths, **two and a half** wavelengths etc., the waves arrive **out of phase** and you get **destructive interference**.

Constructive interference occurs when: path difference $= n\lambda$ (where n is an integer)

Destructive interference occurs when: path difference $= \dfrac{(2n+1)\lambda}{2} = (n + ½)\lambda$

Practice Questions

Q1 Why does the principle of superposition deal with the **vector** sum of two displacements?

Q2 What happens when a crest meets a slightly smaller trough?

Q3 If two points on a wave have a phase difference of 1440°, are they in phase?

Exam Questions

Q1 (a) Two sources are coherent.
What can you say about their frequencies, wavelengths and phase difference? [2 marks]

(b) Suggest why you might have difficulty in observing interference patterns in an area affected by two waves from two sources even though the two sources are coherent. [1 mark]

Q2 Two points on an undamped wave are exactly out of phase.

(a) What is the phase difference between them, expressed in degrees? [1 mark]

(b) Compare the displacements and velocities of the two points. [2 marks]

Learn this and you'll be in a super position to pass your exam...

...I'll get my coat.

There are a few really crucial concepts here: a) interference can be constructive or destructive, b) constructive interference happens when the path difference is a whole number of wavelengths, c) the sources must be coherent.

Standing waves are waves that... er... stand still... well, not still exactly... I mean, well... they don't go anywhere... um...

*You get Standing Waves When a **Progressive Wave** is **Reflected** at a **Boundary***

A standing wave is the **superposition** of **two progressive waves** with the **same wavelength**, moving in **opposite directions**.

1) Unlike progressive waves, **no energy** is transmitted by a standing wave.
2) You can demonstrate standing waves by setting up a **driving oscillator** at one end of a **stretched string** with the other end fixed. The wave generated by the oscillator is **reflected** back and forth.
3) For most frequencies the resultant **pattern** is a **jumble**. However, if the oscillator happens to produce an **exact number of waves** in the time it takes for a wave to get to the **end** and **back again**, then the **original** and **reflected** waves **reinforce** each other.
4) At these **"resonant frequencies"** you get a **standing wave** where the **pattern doesn't move** — it just sits there, bobbing up and down. Happy, at peace with the world...

A sitting wave.

*Standing Waves in **Strings** Form **Oscillating "Loops"** Separated by **Nodes***

1) Each particle vibrates at **right angles** to the string. **Nodes** are where the **amplitude** of the vibration is **zero**. **Antinodes** are points of **maximum amplitude**.
2) At resonant frequencies, an **exact number** of **half wavelengths** fits onto the string.

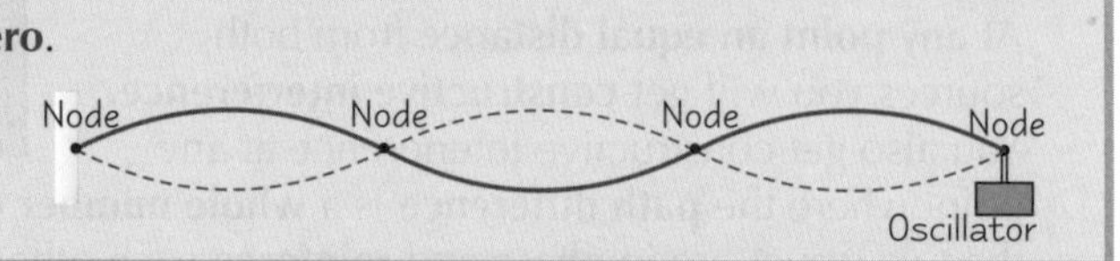

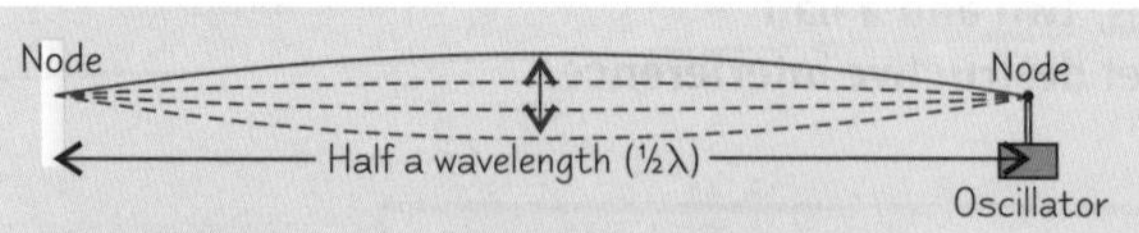

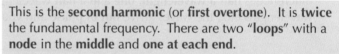

The standing wave above is vibrating at the **lowest possible** resonant frequency (the **fundamental frequency**). It has **one "loop"** with a **node at each end**.

This is the **second harmonic** (or **first overtone**). It is **twice** the fundamental frequency. There are two **"loops"** with a **node** in the **middle** and **one at each end**.

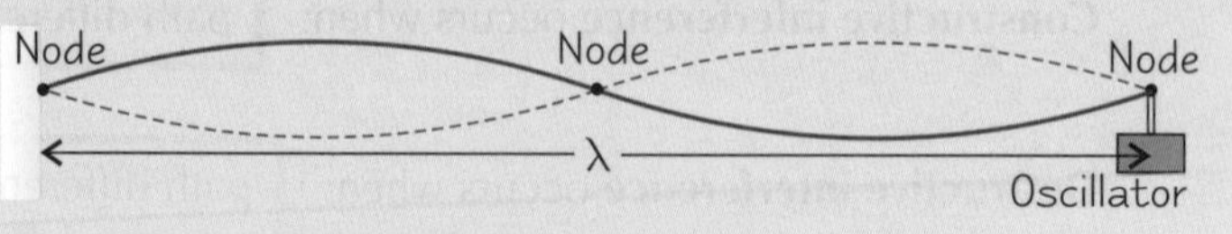

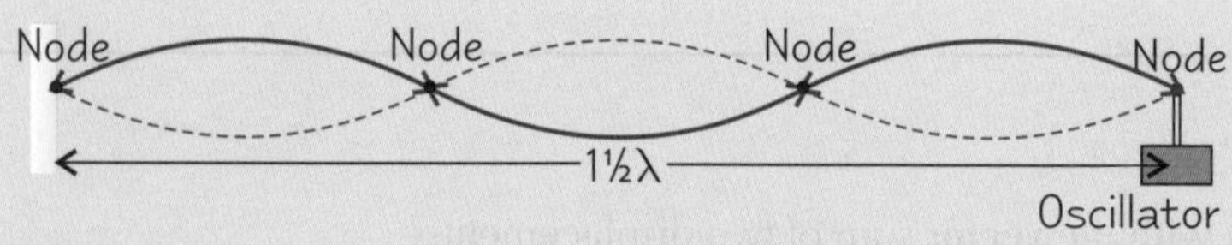

The **third harmonic** (or **second overtone**) is **three times** the fundamental frequency. **1½ wavelengths** fit on the string.

*The **Notes** Played by **Stringed** and **Wind Instruments** are Standing Waves*

Transverse standing waves form on the strings of **stringed instruments** like **violins** and **guitars**. Your finger or the bow sets the **string vibrating** at the point of contact. Waves are sent out in **both directions** and **reflected** back at both ends.

Longitudinal Standing Waves Form in a **Wind Instrument** or Other **Air Column**

1) If a source of sound is placed at the open end of a flute, piccolo, oboe or other column of air, there will be some **frequencies** for which **resonance** occurs and a standing wave is set up.
2) If the instrument has a **closed end**, a **node** will form there. You get the lowest resonant frequency when the length, l, of the pipe is a **quarter wavelength**.

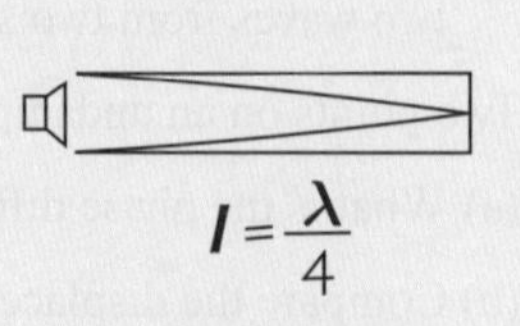

$$l = \frac{\lambda}{4}$$

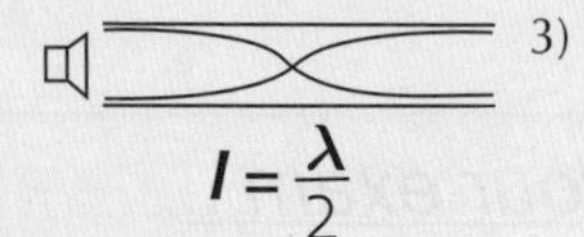

$$l = \frac{\lambda}{2}$$

3) **Antinodes** form at the **open ends** of pipes. If both ends are open, you get the lowest resonant frequency when the length, l, of the pipe is a **half wavelength**.

Remember, the sound waves in wind instruments are <u>longitudinal</u> — they <u>don't</u> actually look like these diagrams.

Standing (Stationary) Waves

You can *Demonstrate Standing Waves* with Microwaves

Microwaves Reflected Off a Metal Plate Set Up a Standing Wave

Microwave standing wave apparatus →

You can find the **nodes** and **antinodes** by moving the **probe** between the **transmitter** and the **reflecting** plate.

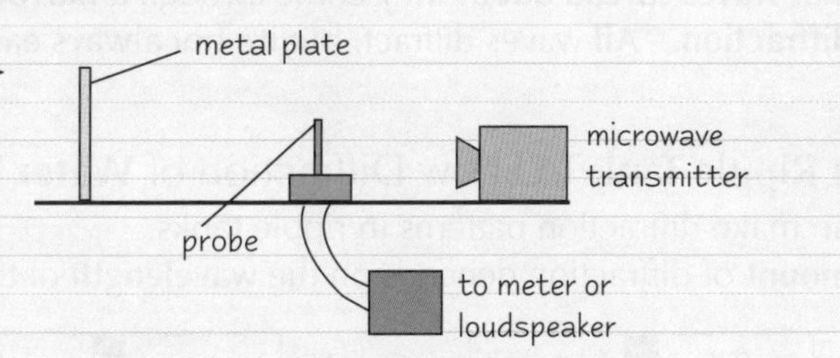

You can Use *Standing Waves* to *Measure* the *Speed of Sound*

Finding the Speed of Sound in a Resonance Tube

1) You can create a closed-end pipe by placing a **hollow tube** into a measuring cylinder of water.
2) Choose a tuning fork and note down the frequency of sound it produces (it'll be stamped on the side of it).
3) Gently tap the tuning fork and hold it just above the hollow tube. The sound waves produced by the fork travel down the tube and get reflected (and form a **node**) at the air/water surface.
4) Move the tube up and down until you find the **shortest distance** between the top of the tube and the water level that the sound from the fork **resonates** at.
5) Just like with any closed pipe, this distance is a **quarter** of the **wavelength** of the standing sound wave.
6) The antinode of the wave actually forms slightly **above** the top of the tube — so you need to add a constant called an **end correction** to the length of your tube **before** you can work out the wavelength.
7) Once you know the **frequency** and **wavelength** of the standing sound wave, you can work out the **speed of sound** (in air), v, using the equation $v = f\lambda$.

Practice Questions

Q1 How do standing waves form?

Q2 At four times the fundamental frequency, how many half wavelengths fit on a violin string?

Q3 Describe an experiment to find the speed of sound in air using standing waves.

Exam Question

Q1 (a) A standing wave of three times the fundamental frequency is formed on a stretched string of length 1.2 m. Sketch a diagram showing the form of the wave. [2 marks]

(b) What is the wavelength of the standing wave? [1 mark]

(c) Explain how the amplitude varies along the string. How is that different from the amplitude of a progressive wave? [2 marks]

CGP — putting the FUN back in FUNdamental frequency...

Resonance was a big problem for the Millennium Bridge in London. The resonant frequency of the bridge was round about normal walking pace, so as soon as people started using it they set up a huge standing wave. An oversight, I feel...

Diffraction

Ripple tanks, ripple tanks — yeah.

*Waves Go **Round Corners** and **Spread out** of **Gaps***

The way that **waves spread out** as they come through a **narrow gap** or go round obstacles is called **diffraction**. **All** waves diffract, but it's not always easy to observe.

Use a **Ripple Tank** To Show Diffraction of **Water Waves**

You can make diffraction patterns in ripple tanks.
The **amount** of diffraction depends on the **wavelength** of the wave compared with the **size of the gap**.

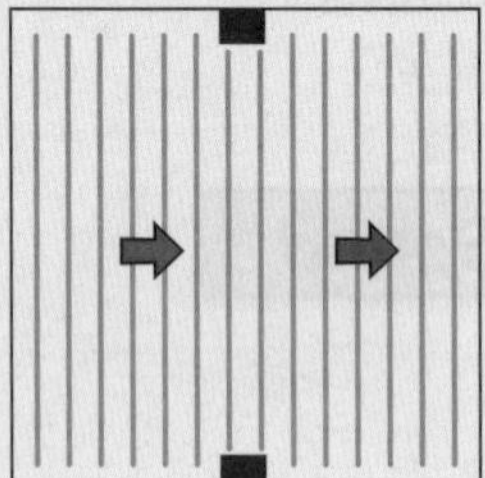

When the gap is **a lot bigger** than the **wavelength**, diffraction is **unnoticeable**.

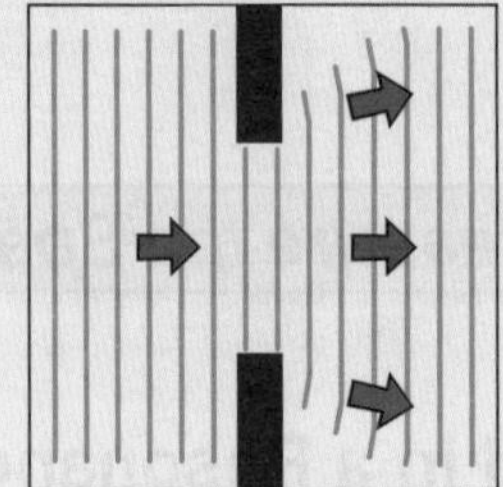

You get **noticeable diffraction** through a gap **several** wavelengths wide.

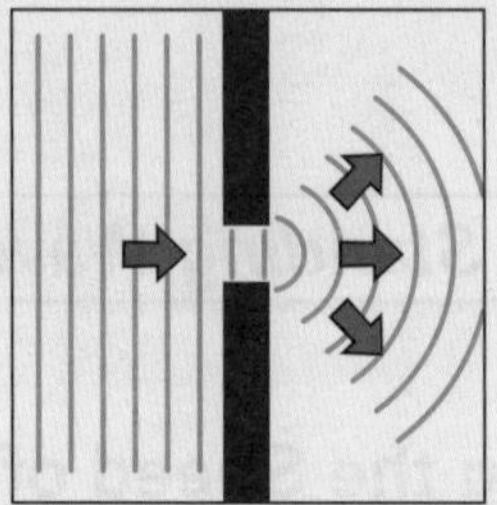

You get the **most** diffraction when the gap is **the same** size as the **wavelength**.

If the gap is **smaller** than the wavelength, the waves are mostly just **reflected back**.

When **sound** passes through a **doorway**, the **size of gap** and the **wavelength** are usually roughly **equal**, so **a lot** of **diffraction** occurs. That's why you have no trouble **hearing** someone through an **open door** to the next room, even if the other person is out of your **line of sight**. The reason that you can't **see** him or her is that when **light** passes through the doorway, it is passing through a **gap** around a **hundred million times bigger** than its wavelength — the amount of diffraction is **tiny**.

Demonstrate **Diffraction** in **Light** Using **Laser Light**

1) Diffraction in **light** can be demonstrated by shining a **laser light** through a very **narrow slit** onto a screen (see page 69). You can alter the amount of diffraction by changing the width of the slit.
2) You can do a similar experiment using a **white light** source instead of the laser (which is monochromatic) and a set of **colour filters**. The size of the slit can be kept constant while the **wavelength** is varied by putting different **colour filters** over the slit.

Warning. Use of coloured filters may result in excessive fun.

*You Get a **Similar** Effect Around an **Obstacle***

When a wave meets an **obstacle**, you get diffraction around the edges.

Behind the obstacle is a '**shadow**', where the wave is blocked. The **wider** the obstacle compared with the wavelength of the wave, the less diffraction you get, and so the **longer** the shadow.

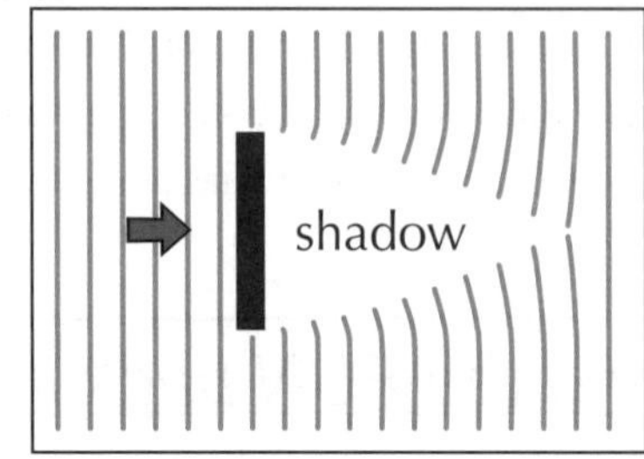

Diffraction

Diffraction is Sometimes **Useful** and Sometimes a Pain...

1) For a **loudspeaker** you want the sound to be heard as widely as possible, so you aim to **maximise** diffraction.
2) With a **microwave oven** you want to **stop** the **microwaves** diffracting out and frying your kidneys **and** you want to **let light through** so you can **see** your food. A **metal mesh** on the **door** has **gaps too small** for microwaves to diffract through, but **light** slips through because of its **tiny wavelength**.

With **Light Waves** you get a **Pattern** of **Light** and **Dark Fringes**

1) If the **wavelength** of a **light wave** is roughly similar to the size of the **aperture**, you get a **diffraction pattern** of light and dark fringes.
2) The pattern has a **bright central fringe** with alternating **dark and bright fringes** on either side of it.

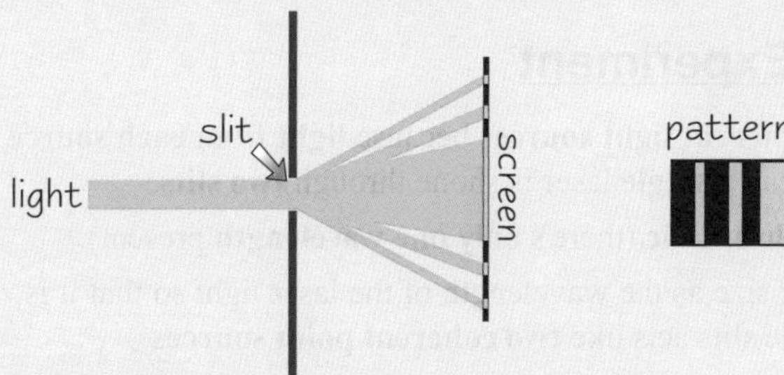

You need to use a coherent light source for this experiment.

3) The **narrower** the slit, the **wider** the diffraction pattern.

You Get a **Similar Pattern** with **Electrons**

1) It's not just with **light** that you get diffraction patterns.
2) In **1927**, two American physicists, **Clinton Davisson** and **Lester Germer**, succeeded in diffracting **electrons**.
3) This was a **huge** discovery. A few years earlier, **Louis de Broglie** had **hypothesised** that electrons would show **wave-like** properties (in the same way that light can show particle-like properties — more about that in the next section), but this was the first **direct evidence** for it.

Electron diffraction patterns look like this

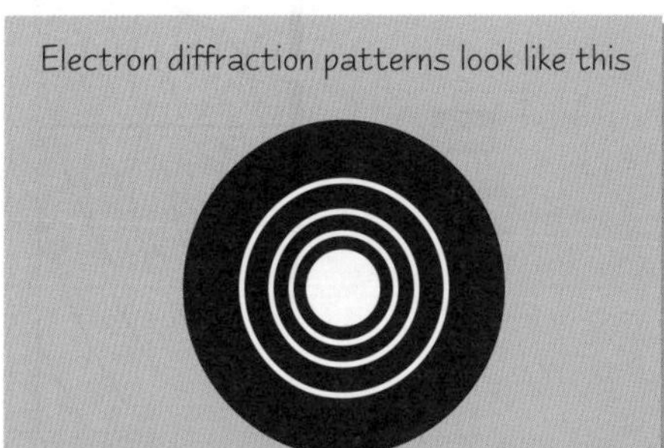

Practice Questions

Q1 What is diffraction?

Q2 Sketch what happens when plane waves meet an obstacle about as wide as one wavelength.

Q3 For a long time some scientists argued that light couldn't be a wave because it did not seem to diffract. Suggest why they might have got this impression.

Q4 Do all waves diffract?

Exam Question

Q1 A mountain lies directly between you and a radio transmitter.
Explain using diagrams why you can pick up long-wave radio broadcasts from the transmitter but not short-wave radio broadcasts. [4 marks]

Even hiding behind a mountain, you can't get away from long-wave radio...

*Diffraction crops up again in particle physics, quantum physics and astronomy, so you **really** need to understand it.*

Two-Source Interference

Yeah, I know, fringe spacing doesn't really sound like a Physics topic — just trust me on this one, OK.

Demonstrating Two-Source Interference in **Water** and **Sound** is Easy

1) It's **easy** to demonstrate **two-source interference** for either **sound** or **water** because they've got **wavelengths** of a handy **size** that you can **measure**.
2) You need **coherent** sources, which means the **wavelength** and **frequency** have to be the **same**. The trick is to use the **same oscillator** to drive **both sources**. For **water**, one **vibrator drives two dippers**. For sound, **one oscillator** is connected to **two loudspeakers**. (See diagram on page 65.)

Demonstrating **Two-Source** Interference for **Light** is Harder

Young's Double-Slit Experiment

1) You **can't** arrange **two separate coherent light sources** because **light** from **each source** is emitted in **random bursts**. Instead a **single** laser is shone through **two slits**.
2) Laser light is **coherent** and **monochromatic** (there's only **one wavelength** present).
3) The slits have to be about the same size as the wavelength of the laser light so that it is **diffracted** — then the light from the slits acts like **two coherent point sources**.

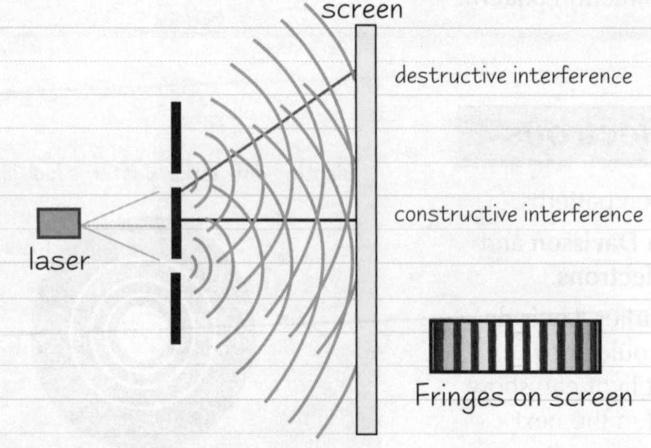

4) You get a pattern of light and dark **fringes**, depending on whether constructive or destructive **interference** is taking place. Thomas Young — the first person to do this experiment (with a lamp rather than a laser) — came up with an **equation** to **work out** the **wavelength** of the **light** from this experiment (see next page).

You Can Do a **Similar** Experiment with **Microwaves**

1) To see interference patterns with **microwaves**, you can **replace** the laser and slits with two microwave **transmitter cones** attached to the **same** signal generator.
2) You also need to replace the screen with a microwave **receiver probe** (like the one used in the standing waves experiment on page 67).
3) If you move the probe along the path of the green arrow, you'll get an **alternating pattern** of **strong** and **weak** signals — just like the light and dark fringes on the screen.

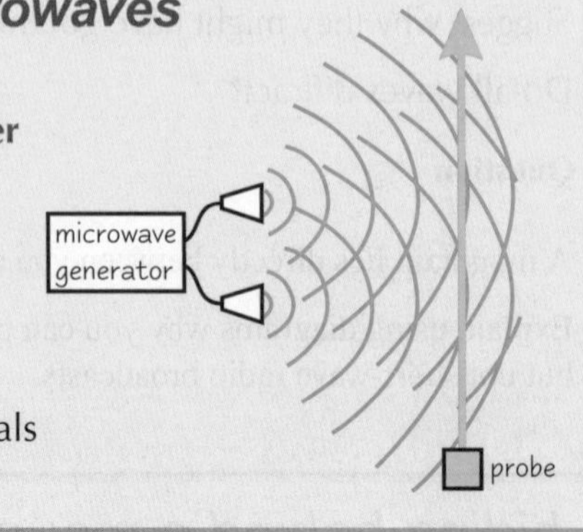

Two-Source Interference

Work Out the Wavelength with **Young's Double-Slit Formula**

1) The fringe spacing (***X***), wavelength (***λ***), spacing between slits (***d***) and the distance from slits to screen (***D***) are all related by **Young's double-slit formula**, which works for all waves (you need to know it, but not derive it).

Fringe spacing, $X = \frac{D\lambda}{d}$

"Fringe spacing" means the distance from the centre of one minimum to the centre of the next minimum or from the centre of one maximum to the centre of the next maximum.

2) Since the wavelength of light is so small you can see from the formula that a high ratio of ***D / d*** is needed to make the fringe spacing **big enough to see**.
3) Rearranging, you can use **$\lambda = Xd / D$** to **calculate the wavelength** of light.
4) The fringes are **so tiny** that it's very hard to get an **accurate value of X**. It's easier to measure across **several** fringes then **divide** by the number of **fringe widths** between them.

Always check your fringe spacing.

Young's Experiment was ***Evidence*** for the ***Wave Nature*** of Light

1) Towards the end of the **17th century**, two important **theories of light** were published — one by Isaac Newton and the other by a chap called Huygens. **Newton's** theory suggested that light was made up of tiny particles, which he called "**corpuscles**". And **Huygens** put forward a theory using **waves**.
2) The **corpuscular theory** could explain **reflection** and **refraction**, but **diffraction** and **interference** are both **uniquely** wave properties. If it could be **shown** that light showed interference patterns, that would help settle the argument once and for all.
3) **Young's** double-slit experiment (over 100 years later) provided the necessary evidence. It showed that light could both **diffract** (through the narrow slits) and **interfere** (to form the interference pattern on the screen).

Of course, this being Physics, nothing's ever simple — give it another 100 years or so and the debate would be raging again. But that can wait for the next section...

Practice Questions

Q1 In Young's experiment, why do you get a bright fringe at a point equidistant from both slits?

Q2 What does Young's experiment show about the nature of light?

Q3 Write down Young's double-slit formula.

Exam Questions

Q1 (a) The diagram on the right shows waves from two coherent light sources, S_1 and S_2. Sketch the interference pattern, marking on constructive and destructive interference. [2 marks]

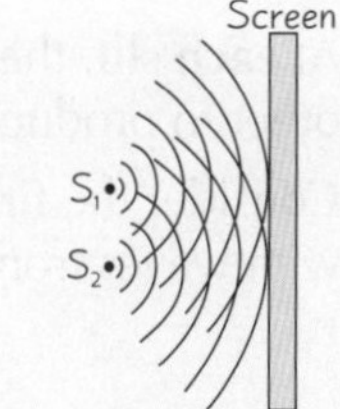

(b) In practice if interference is to be observed, S_1 and S_2 must be slits in a screen behind which there is a source of laser light. Why? [2 marks]

Q2 In an experiment to study sound interference, two loudspeakers are connected to an oscillator emitting sound at 1320 Hz and set up as shown in the diagram below. They are 1.5 m apart and 7 m away from the line AC. A listener moving from A to C hears minimum sound at A and C and maximum sound at B.

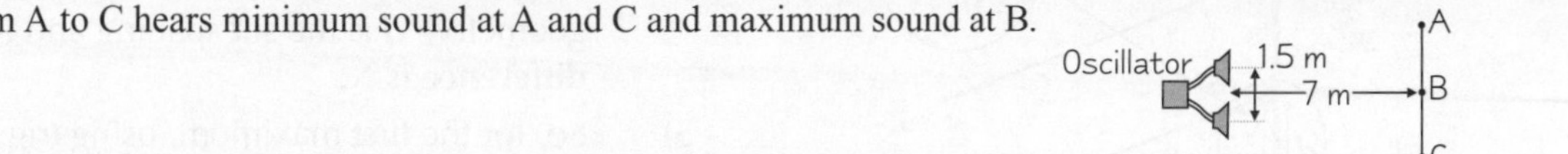

(a) Calculate the wavelength of the sound waves if the speed of sound in air is taken to be 330 ms^{-1}. [1 mark]

(b) Calculate the separation of points A and C. [2 marks]

Carry on Physics — this page is far too saucy...

Be careful when you're calculating the fringe width by averaging over several fringes. Don't just divide by the number of bright lines. Ten bright lines will only have nine fringe-widths between them, not ten. It's an easy mistake to make, but you have been warned... mwa ha ha ha (felt necessary, sorry).

*

* *Sadly, Barbara Windsor was unavailable.*

Diffraction Gratings

Ay... starting to get into some pretty funky stuff now. I like light experiments.

*Interference Patterns Get **Sharper** When You Diffract Through **More Slits***

1) You can repeat **Young's double-slit** experiment (see p. 70) with **more than two equally spaced** slits. You get basically the **same shaped** pattern as for two slits — but the **bright bands** are **brighter** and **narrower** and the **dark areas** between are **darker**.
2) When **monochromatic light** (one wavelength) is passed through a **grating** with **hundreds** of slits per millimetre, the interference pattern is **really sharp** because there are so **many beams reinforcing** the **pattern**.
3) Sharper fringes make for more **accurate** measurements.

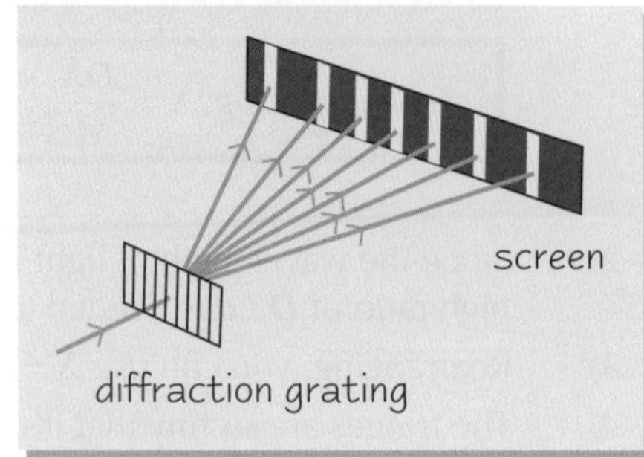

Monochromatic Light** on a **Diffraction Grating** gives **Sharp Lines

1) For **monochromatic** light, all the **maxima** are sharp lines. (It's different for white light — see next page.)
2) There's a line of **maximum brightness** at the centre called the **zero order** line.
3) The lines just **either side** of the central one are called **first order lines**. The **next pair out** are called **second order** lines and so on.
4) For a grating with slits a distance d apart, the angle between the **incident beam** and **the nth order maximum** is given by:

$$d \sin \theta = n\lambda$$

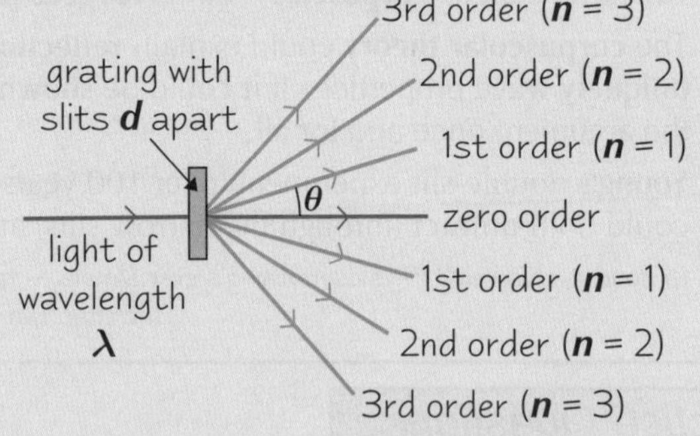

5) So by observing d, θ and n you can **calculate the wavelength** of the light.

If the grating has N slits per metre, then the slit spacing, d, is just 1/N metres.

WHERE THE EQUATION COMES FROM:

1) At **each slit**, the incoming waves are **diffracted**. These diffracted waves then **interfere** with each other to produce an **interference pattern.**
2) Consider the **first order maximum**. This happens at the **angle** when the waves from one slit line up with waves from the **next slit** that are **exactly one wavelength** behind.

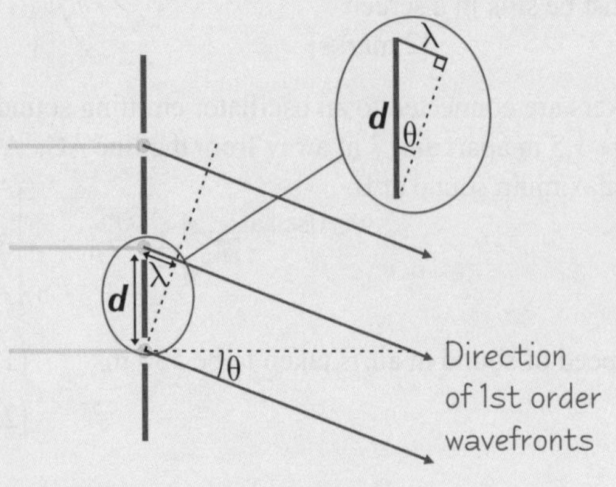

3) Call the **angle** between the **first order maximum** and the **incoming light** θ.
4) Now, look at the **triangle** highlighted in the diagram. The angle is θ (using basic geometry), d is the slit spacing and the **path difference** is λ.
5) So, for the first maximum, using trig:

$$d \sin \theta = \lambda$$

6) The other maxima occur when the path difference is 2λ, 3λ, 4λ, etc. So to make the equation **general**, just replace λ with $n\lambda$, where n is an integer — the **order** of the maximum.

Diffraction Gratings

You can Draw **General Conclusions** from **d** $\sin\theta = n\lambda$

1) If λ is **bigger**, $\sin\theta$ is **bigger**, and so θ is **bigger**. This means that the larger the **wavelength**, the more the pattern will **spread out**.
2) If d is **bigger**, $\sin\theta$ is **smaller**. This means that the **coarser** the **grating**, the **less** the pattern will **spread out**.
3) Values of $\sin\theta$ greater than **1** are **impossible**. So if for a certain n you get a result of **more than 1** for $\sin\theta$ you know that that order **doesn't exist**.

Shining **White Light** Through a **Diffraction Grating** Produces **Spectra**

1) **White light** is really a **mixture** of **colours**. If you **diffract** white light through a **grating** then the patterns due to **different wavelengths** within the white light are **spread out** by **different** amounts.
2) Each **order** in the pattern becomes a **spectrum**, with **red** on the **outside** and **violet** on the **inside**. The **zero order maximum** stays **white** because all the wavelengths just pass straight through.

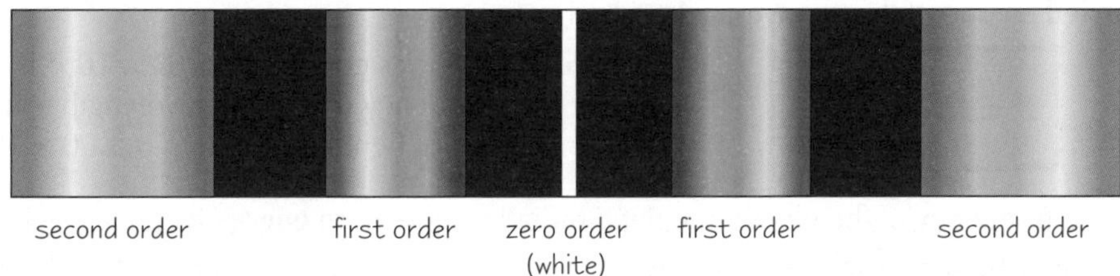

Astronomers and **chemists** often need to study spectra to help identify elements. They use diffraction gratings rather than prisms because they're **more accurate**.

Practice Questions

Q1 How is the diffraction grating pattern for white light different from the pattern for laser light?

Q2 What difference does it make to the pattern if you use a finer grating?

Q3 What equation is used to find the angle between the nth order maximum and the incident beam for a diffraction grating?

Exam Questions

Q1 Yellow laser light of wavelength 600 nm (6×10^{-7} m) is transmitted through a diffraction grating of 4×10^5 lines per metre.

(a) At what angle to the normal are the first and second order bright lines seen? [4 marks]

(b) Is there a fifth order line? [1 mark]

Q2 Visible, monochromatic light is transmitted through a diffraction grating of 3.7×10^5 lines per metre. The first order maximum is at an angle of 14.2° to the incident beam.

Find the wavelength of the incident light. [2 marks]

Ooooooooooooo — pretty patterns...

Yes, it's the end of another beautiful section — woohoo. Three important points for you to take away — the more slits you have, the sharper the image, one lovely equation to learn and white light makes a pretty spectrum. Make sure you get everything in this section — there's some good stuff coming up in the next one and I wouldn't want you to be distracted.

Light — Wave or Particle

You probably already thought light was a bit weird — but oh no... being a wave that travels at the fastest speed possible isn't enough for light — it has to go one step further and act like a particle too...

Light Behaves Like a Wave... or a Stream of Particles

1) In the **late nineteenth century**, if you asked what light was, scientists would happily show you lots of nice experiments showing how light must be a **wave** (see Unit 2: Section 2).
2) Then came the **photoelectric effect** (p. 76), which mucked up everything. The only way you could explain this was if light acted as a **particle** — called a **photon**.

A Photon is a Quantum of EM Radiation

1) When Max Planck was investigating **black body radiation** (don't worry — you don't need to know about that just yet), he suggested that **EM waves** can **only** be **released** in **discrete packets**, called **quanta.** A single packet of **EM radiation** is called a **quantum**.

   The **energy carried** by one of these **wave-packets** had to be:

$$E = hf = \frac{hc}{\lambda}$$

where h = Planck's constant = 6.63×10^{-34} Js, f = frequency (Hz), λ = wavelength (m) and c = speed of light in a vacuum = 3.00×10^{8} ms^{-1}

2) So, the **higher** the **frequency** of the electromagnetic radiation, the more **energy** its wave-packets carry.
3) **Einstein** went **further** by suggesting that **EM waves** (and the energy they carry) can only **exist** in discrete packets. He called these wave-packets **photons**.
4) He believed that a photon acts as **particle**, and will either transfer **all** or **none** of its energy when interacting with another particle, like an electron.

Photon Energies are Usually Given in Electronvolts

1) The **energies involved** when you're talking about photons are **so tiny** that it makes sense to use a more **appropriate unit** than the **joule**. Bring on the **electronvolt** ...
2) When you **accelerate** an electron between two electrodes, it transfers some electrical potential energy (eV) into kinetic energy.

$$eV = \frac{1}{2}mv^2$$

e is the charge on an electron: 1.6×10^{-19} C.

3) An electronvolt is defined as:

   The **kinetic energy gained** by an **electron** when it is **accelerated** through a **potential difference** of **1 volt**.

4) So 1 electron volt = $e \times V = 1.6 \times 10^{-19}$ C × 1 JC^{-1}. ⟹ $1 \text{ eV} = 1.6 \times 10^{-19}$ J

Threshold Voltage is Used to Find Planck's Constant

1) Planck's constant comes up everywhere — but it's not just some random number plucked out of the air. You can find its value by doing a simple experiment with **light-emitting diodes** (**LED**s).
2) Current will only pass through an LED after a **minimum voltage** is placed across it — the **threshold voltage** V_0.
3) This is the voltage needed to give the electrons the **same energy** as a photon emitted by the LED. **All** of the electron's **kinetic energy** after it is accelerated over this potential difference is **transferred** into a **photon**.
4) So by finding the threshold voltage for a particular wavelength LED, you can calculate Planck's constant.

$$E = \frac{hc}{\lambda} = eV_0 \Rightarrow h = \frac{(eV_0)\lambda}{c}$$

Light — Wave or Particle

You can Use LEDs to Calculate Planck's Constant

You've just seen the **theory** of how to find **Planck's constant** — now it's time for the **practicalities**.

Experiment to Measure Planck's Constant

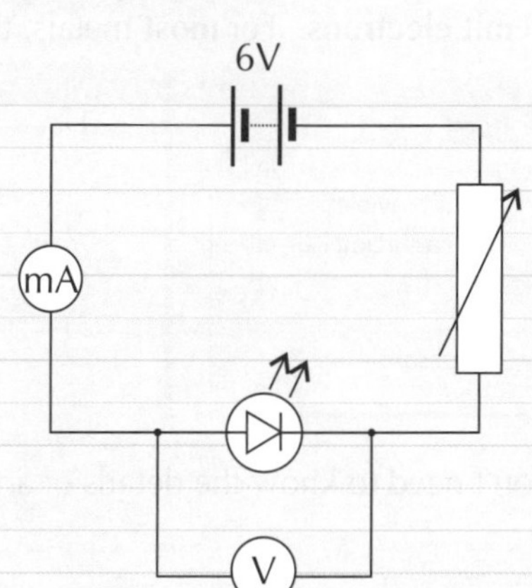

1) Connect an LED of known wavelength in the electrical circuit shown.
2) Start off with no current flowing through the circuit, then adjust the variable resistor until a current just begins to flow through the circuit.
3) Record the voltage (V_0) across the LED, and the wavelength of light the LED emits.
4) Repeat this experiment with a number of LEDs that emit different optical wavelengths.

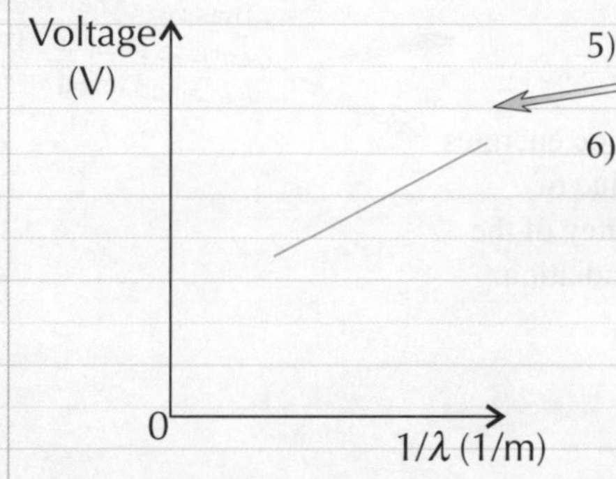

5) Plot a graph of threshold voltages (V_0) against $1/\lambda$ (where λ is the wavelength of light emitted by the LED in metres).
6) You should get a straight line graph with a gradient of hc/e — which you can then use to find the value of h.

E.g. $$\textbf{gradient} = \frac{hc}{e} = 1.24 \times 10^{-6},$$

$$\text{so } h = \frac{1.24 \times 10^{-6} e}{c} = \frac{(1.24 \times 10^{-6}) \times (1.6 \times 10^{-19})}{3 \times 10^{8}}$$

$$= \mathbf{6.6 \times 10^{-34}\ Js}\ (2\text{ s.f.})$$

Practice Questions

Q1 Give two different ways to describe the nature of light.

Q2 Write down the two formulas you can use to find the energy of a photon. Include the meanings of all the symbols you use.

Q3 What is an electronvolt? What is 1 eV in joules?

Exam Question

$c = 3.00 \times 10^{8}\ ms^{-1}$

Q1 An LED is tested and found to have a threshold voltage of 1.70 V.

(a) Find the energy of the photons emitted by the LED. Give your answer in joules. [2 marks]

(b) The LED emits light with a wavelength of 700 nm.
Use your answer from a) to calculate the value of Planck's constant. [2 marks]

Millions of light particles are hitting your retinas as you read this... PANIC...

I hate it in physics when they tell you lies, make you learn it, and just when you've got to grips with it they tell you it was all a load of codswallop. This is the real deal folks — light isn't just the nice wave you've always known...

The Photoelectric Effect

The photoelectric effect was one of the original troublemakers in the light-is-it-a-wave-or-a-particle problem...

Shining Light on a Metal can Release Electrons

If you shine **light** of a **high enough frequency** onto the **surface of a metal**, it will **emit electrons**. For **most** metals, this **frequency** falls in the **U.V.** range.

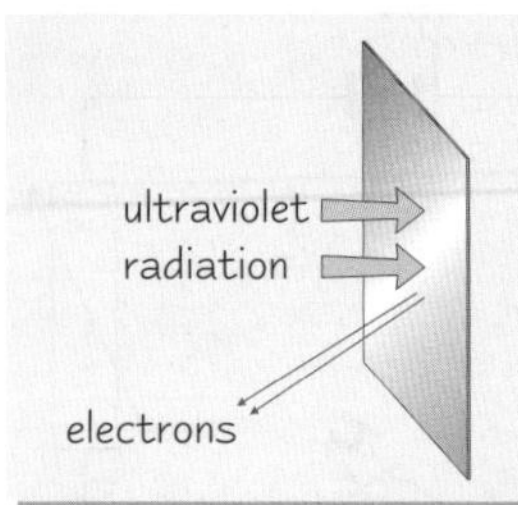

1) **Free electrons** on the **surface** of the metal **absorb energy** from the light, making them **vibrate**.
2) If an electron **absorbs enough** energy, the **bonds** holding it to the metal **break** and the electron is **released**.
3) This is called the **photoelectric effect** and the electrons emitted are called **photoelectrons**.

You don't need to know the details of any experiments on this — you just need to learn the three main conclusions:

Conclusion 1	For a given metal, **no photoelectrons are emitted** if the radiation has a frequency **below** a certain value — called the **threshold frequency**.
Conclusion 2	The photoelectrons are emitted with a variety of kinetic energies ranging from zero to some maximum value. This value of **maximum kinetic energy** increases with the **frequency** of the radiation, and is **unaffected** by the **intensity** of the radiation.
Conclusion 3	The **number** of photoelectrons emitted per second is **proportional** to the **intensity** of the radiation.

These are the two that had scientists puzzled. They can't be explained using wave theory.

The Photoelectric Effect Couldn't be Explained by Wave Theory

According to wave theory:
1) For a particular frequency of light, the **energy** carried is **proportional** to the **intensity** of the beam.
2) The energy carried by the light would be **spread evenly** over the wavefront.
3) **Each** free electron on the surface of the metal would gain a **bit of energy** from each incoming wave.
4) Gradually, each electron would gain **enough energy** to leave the metal.

SO... If the light had a **lower frequency** (i.e. was carrying less energy) it would take **longer** for the electrons to gain enough energy — but it would happen eventually. There is **no explanation** for the **threshold frequency**.

The **higher the intensity** of the wave, the **more energy** it should transfer to each electron — the kinetic energy should increase with **intensity**. There's **no explanation** for the **kinetic energy** depending only on the **frequency**.

The Photon Model Explained the Photoelectric Effect Nicely

According to the photon model (see page 74)**:**
1) When light hits its surface, the metal is **bombarded** by photons.
2) If one of these photons **collides** with a free electron, the electron will gain energy equal to ***hf***.

Before an electron can **leave** the surface of the metal, it needs enough energy to **break the bonds holding it there**. This energy is called the **work function energy** (symbol ϕ, phi) and its **value** depends on the **metal**.

The Photoelectric Effect

It Explains the Threshold Frequency...

1) If the energy **gained** from the photon is **greater** than the **work function energy**, the electron can be **emitted.**
2) If it **isn't**, the electron will just **shake about a bit**, then release the energy as another photon. The metal will heat up, but **no electrons** will be emitted.
3) Since for **electrons** to be released, $hf \geq \phi$, the **threshold frequency** must be:

$$f = \frac{\phi}{h}$$

In theory, if a second photon hit an electron before it released the energy from the first, it could gain enough to leave the metal. This would have to happen very quickly though. An electron releases any excess energy after about 10^{-8} s. That's 0.000 000 01 s — safe to say, the chances of that happening are pretty slim.

... and the Maximum Kinetic Energy

1) The **energy transferred** to an electron is **h*f*.**
2) The **kinetic energy** it will be carrying when it **leaves** the metal will be h*f* **minus** any energy it's **lost** on the way out (there are loads of ways it can do that, which explains the **range** of energies).
3) The **minimum** amount of energy it can lose is the **work function energy**, so the **maximum kinetic energy** is given by the equation:

$$hf = \phi + \frac{1}{2}mv_{max}^2$$

4) The **kinetic energy** of the electrons is **independent of the intensity**, because they can **only absorb one photon** at a time.

Practice Questions

Q1 Describe an experiment that demonstrates the photoelectric effect.

Q2 What is meant by the threshold frequency?

Q3 Write down the equation that relates the work function of a metal and the threshold frequency.

Q4 Write an equation that relates the maximum kinetic energy of a photoelectron released from a metal surface and the frequency of the incident light on the surface.

Exam Questions

$h = 6.63 \times 10^{-34}$ Js

Q1 The work function of calcium is 2.9 eV.
Find the threshold frequency of radiation needed for the photoelectric effect to take place. [2 marks]

Q2 The surface of a copper plate is illuminated with monochromatic ultraviolet light, with a frequency of 2.0×10^{15} Hz. The work function for copper is 4.7 eV.
(a) Find the energy in eV carried by one ultraviolet photon. [3 marks]
(b) Find the maximum kinetic energy of a photoelectron emitted from the copper surface. [2 marks]

Q3 Explain why the photoelectric effect only occurs after the incident light has reached a certain frequency. [2 marks]

I'm so glad we got that all cleared up...

Well, that's about as hard as it gets at AS. The most important bits here are why wave theory doesn't explain the phenomenon, and why the photon theory does. A good way to learn conceptual stuff like this is to try to explain it to someone else. You'll get the formulas in your handy data book, but it's probably a good idea to learn them too...

Energy Levels and Photon Emission

Electrons in Atoms Exist in Discrete Energy Levels

1) **Electrons** in an **atom** can **only exist** in certain **well-defined energy levels**. Each level is given a **number**, with **n = 1** representing the **ground state**.
2) Electrons can **move down** an energy level by **emitting** a **photon**.
3) Since these **transitions** are between **definite energy levels**, the **energy** of **each photon** emitted can **only** take a **certain allowed value**.
4) The diagram on the right shows the **energy levels** for **atomic hydrogen**.
5) On the diagram, energies are labelled in both **joules** and **electonvolts** for **comparison's** sake.
6) The **energy** carried by each **photon** is **equal** to the **difference in energies** between the **two levels**. The equation below shows a **transition** between levels **n = 2** and **n = 1**:

$$\Delta E = E_2 - E_1 = hf = \frac{hc}{\lambda}$$

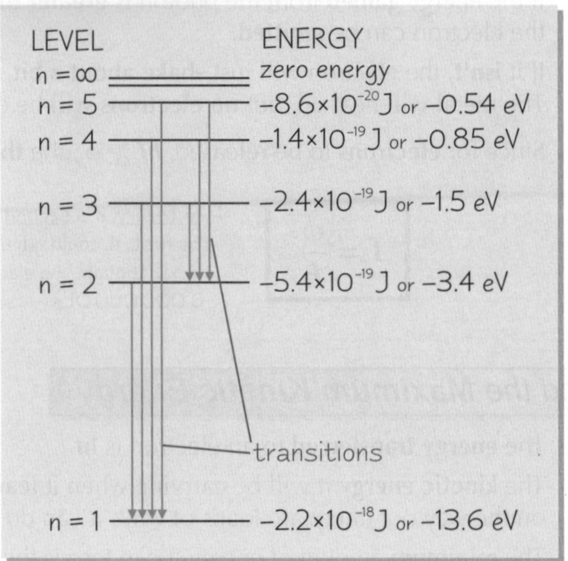

The energies are only negative because of how "zero energy" is defined. Just one of those silly convention things — don't worry about it.

Hot Gases Produce Line Emission Spectra

1) If you heat a gas to a high temperature, many of it's electrons move to higher energy levels.
2) As they fall back to the ground state, these electrons emit energy as photons.
3) If you **split** the light from a **hot gas** with a **prism** or a **diffraction grating** (see pages 72-73), you get a **line spectrum**. A line spectrum is seen as a **series** of **bright lines** against a **black background**, as shown below.

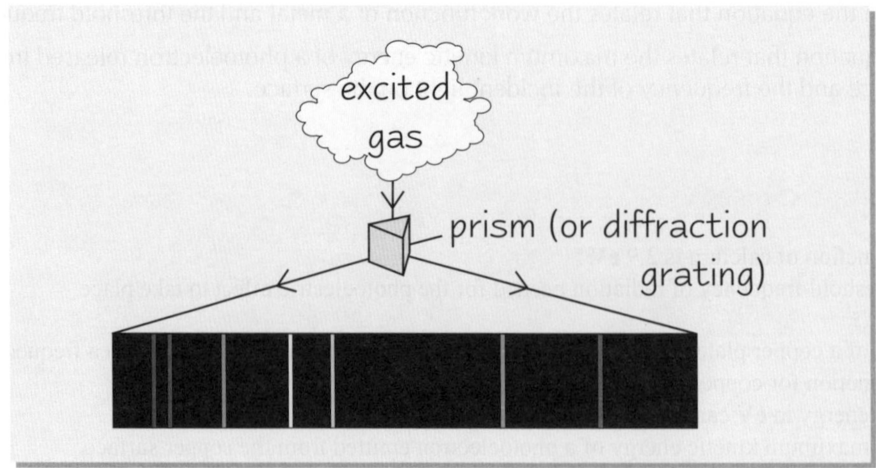

4) Each **line** on the spectrum corresponds to a **particular wavelength** of light **emitted** by the source. Since only **certain photon energies** are **allowed**, you only see the **corresponding wavelengths**.

Energy Levels and Photon Emission

Shining **White Light** through a **Cool Gas** gives an **Absorption Spectrum**

Continuous Spectra Contain **All** Possible **Wavelengths**

1) The **spectrum** of **white light** is **continuous**.
2) If you **split** the **light** up with a **prism**, the **colours** all **merge** into each other — there **aren't** any **gaps** in the spectrum.
3) **Hot things** emit a **continuous spectrum** in the visible and infrared.

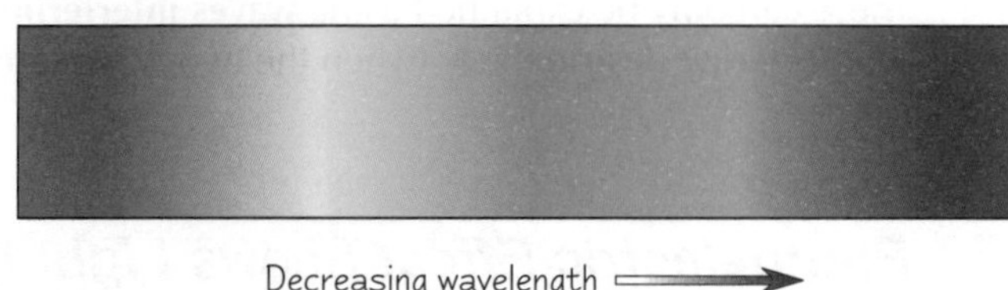

Cool Gases **Remove** Certain **Wavelengths** from the Continuous Spectrum

1) You get a **line absorption spectrum** when **light** with a **continuous spectrum** of **energy** (white light) passes through a cool gas.
2) At **low temperatures**, **most** of the **electrons** in the **gas atoms** will be in their **ground states**.
3) **Photons** of the **correct wavelength** are **absorbed** by the **electrons** to **excite** them to **higher energy levels**.
4) These **wavelengths** are then **missing** from the **continuous spectrum** when it **comes out** the other side of the gas.
5) You see a **continuous spectrum** with **black lines** in it corresponding to the **absorbed wavelengths**.
6) If you **compare** the **absorption** and **emission spectra** of a **particular gas**, the **black lines** in the **absorption spectrum match up** to the **bright lines** in the **emission spectrum**.

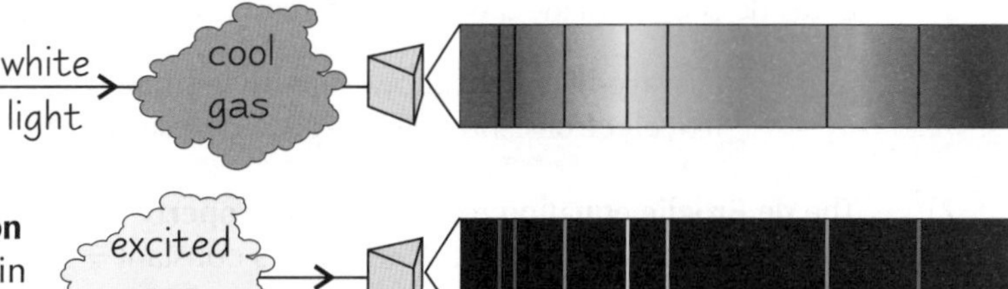

Practice Questions

Q1 Describe line absorption and line emission spectra. How are these two types of spectra produced?

Q2 Use the size of the energy level transitions involved to explain how the coating on a fluorescent tube converts UV into visible light.

Exam Question

$e = 1.6 \times 10^{-19}$ C

Q1 An electron is accelerated through a potential difference of 12.1 V.

(a) How much kinetic energy has it gained in (i) eV and (ii) joules? [2 marks]

(b) This electron hits a hydrogen atom and excites it.

(i) Explain what is meant by excitation. [1 mark]

(ii) Using the energy values on the right, work out to which energy level the electron from the hydrogen atom is excited. [1 mark]

(iii) Calculate the energies of the three photons that might be emitted as the electron returns to its ground state. [3 marks]

n = 5 — – 0.54 eV
n = 4 — – 0.85 eV
n = 3 — – 1.5 eV
n = 2 — – 3.4 eV
n = 1 — – 13.6 eV

I can honestly say I've never got so excited that I've produced light...

This is heavy stuff, it really is. Quite interesting though, as I was just saying to Dom a moment ago. He's doing a psychology book. Psychology's probably quite interesting too — and easier. But it won't help you become an astrophysicist.

Is it a wave? Is it a particle? No, it's a wave. No, it's a particle. No it's not, it's a wave. No don't be daft, it's a particle. (etc.)

Interference and **Diffraction** show **Light** as a **Wave**

1) Light produces **interference** and **diffraction** patterns — **alternating bands** of **dark** and **light**.
2) These can **only** be explained using **waves interfering constructively** (when two waves overlap in phase) or **interfering destructively** (when the two waves are out of phase). (See p.65.)

The **Photoelectric Effect** Shows **Light** Behaving as a **Particle**

1) **Einstein** explained the results of **photoelectricity experiments** (see p.76) by thinking of the **beam of light** as a series of **particle-like photons**.
2) If a **photon** of light is a **discrete** bundle of energy, then it can **interact** with an **electron** in a **one-to-one way**.
3) **All** the **energy** in the **photon** is **given** to one **electron**.

De Broglie Came up With the **Wave-Particle Duality Theory**

1) Louis de Broglie made a **bold suggestion** in his **PhD thesis**:

   If **'wave-like' light** showed **particle properties** (photons), **'particles'** like **electrons** should be expected to show **wave-like properties**.

2) The **de Broglie equation** relates a **wave property** (**wavelength**, λ) to a **moving particle property** (**momentum**, mv). h = Planck's constant = 6.63×10^{-34} Js.

$$\lambda = \frac{h}{mv}$$

3) The **de Broglie wave** of a particle can be interpreted as a **'probability wave'**.
   (The probability of finding a particle at a point is directly proportional to the square of the amplitude of the wave at that point — but you don't need to know that for your exam.)
4) Many physicists at the time **weren't very impressed** — his ideas were just **speculation**. But later experiments **confirmed** the wave nature of electrons.

I'm not impressed — this is just speculation. What do you think Dad?

Electron Diffraction shows the **Wave Nature** of **Electrons**

1) **Diffraction patterns** are observed when **accelerated electrons** in a vacuum tube **interact** with the **spaces** in a graphite **crystal**.
2) This **confirms** that electrons show **wave-like** properties.
3) According to wave theory, the **spread** of the **lines** in the diffraction pattern **increases** if the **wavelength** of the wave is **greater**.
4) In electron diffraction experiments, a **smaller accelerating voltage**, i.e. **slower** electrons, gives **widely spaced** rings.
5) **Increase** the **electron speed** and the diffraction pattern circles **squash together** towards the **middle**. This fits in with the **de Broglie** equation above — if the **velocity** is **higher**, the **wavelength** is **shorter** and the **spread** of lines is **smaller**.

In general, λ for **electrons** accelerated in a **vacuum tube** is about the **same size** as **electromagnetic waves** in the **X-ray** part of the spectrum.

Wave-Particle Duality

Particles Don't show Wave-Like Properties All the Time

You **only** get **diffraction** if a particle interacts with an object of about the **same size** as its **de Broglie wavelength**.

A **tennis ball**, for example, with **mass 0.058 kg** and **speed 100 ms^{-1}** has a **de Broglie wavelength** of **10^{-34} m**. That's **10^{19} times smaller** than the **nucleus** of an **atom**! There's nothing that small for it to interact with.

Example An electron of mass 9.11×10^{-31} kg is fired from an electron gun at 7×10^{6} ms^{-1}. What size object will the electron need to interact with in order to diffract?

Momentum of electron = $mv = 6.38 \times 10^{-24}$ kg ms^{-1}

$\lambda = h/mv = 6.63 \times 10^{-34} / 6.38 \times 10^{-24} = 1 \times 10^{-10}$ m

Only crystals with atom layer spacing around this size are likely to cause the diffraction of this electron.

A **shorter wavelength** gives **less diffraction effects**. This fact is used in the **electron microscope**.

Diffraction effects **blur detail** on an image. If you want to **resolve tiny detail** in an **image**, you need a **shorter wavelength**. **Light** blurs out detail more than '**electron-waves**' do, so an **electron microscope** can resolve **finer detail** than a **light microscope**. They can let you look at things as tiny as a single strand of DNA... which is nice.

Practice Questions

Q1 Which observations show light to have a 'wave-like' character?

Q2 Which observations show light to have a 'particle' character?

Q3 What happens to the de Broglie wavelength of a particle if its velocity increases?

Q4 Which observations show electrons to have a 'wave-like' character?

Exam Questions

$h = 6.63 \times 10^{-34}$ Js ; $c = 3.00 \times 10^{8}$ ms^{-1} ; electron mass = 9.11×10^{-31} kg ; proton mass = $1840 \times$ electron mass

Q1 (a) State what is meant by the wave-particle duality of electromagnetic radiation. [1 mark]

(b) (i) Calculate the energy in joules and in electronvolts of a photon of wavelength 590 nm. [3 marks]

(ii) Calculate the speed of an electron which will have the same wavelength as the photon in (b)(i). [2 marks]

Q2 Electrons travelling at a speed of 3.5×10^{6} ms^{-1} exhibit wave properties.

(a) Calculate the wavelength of these electrons. [2 marks]

(b) Calculate the speed of protons which would have the same wavelength as these electrons. [2 marks]

(c) Both electrons and protons were accelerated from rest by the same potential difference.
Explain why they will have different wavelengths.
(Hint: if they're accelerated by the same p.d., they have the same K.E.) [3 marks]

Q3 An electron is accelerated through a potential difference of 6.0 kV.

(a) Calculate its kinetic energy in joules, assuming no energy is lost in the process. [2 marks]

(b) Using the data above, calculate the speed of the electron. [2 marks]

(c) Calculate the de Broglie wavelength of the electron. [2 marks]

Don't hide your wave-particles under a bushel...

Right — I think we'll all agree that quantum physics is a wee bit strange when you come to think about it. What it's saying is that electrons and photons aren't really waves, and they aren't really particles — they're ***both****... at the* ***same time****. It's what quantum physicists like to call a 'juxtaposition of states'. Well they would, wouldn't they...*

Science is all about getting good evidence to test your theories... and part of that is knowing how good the results from an experiment are. Physicists always have to include the uncertainty in a result, so you can see the range the actual value probably lies within. Dealing with error and uncertainty is an important skill, so those pesky examiners like to sneak in a couple of questions about it... but if you know your stuff you can get some easy marks.

Nothing is Certain

1) **Every** measurement you take has an **experimental uncertainty**. Say you've done something outrageous like measure the length of a piece of wire with a centimetre ruler. You might think you've measured its length as 30 cm, but at **best** you've probably measured it to be 30 ± **0.5** cm. And that's without taking into account any other errors that might be in your measurement...
2) The ± bit gives you the **range** in which the **true** length (the one you'd really like to know) probably lies — 30 ± 0.5 cm tells you the true length is very likely to lie in the range of 29.5 to 30.5 cm.
3) The smaller the uncertainty, the nearer your value must be to the true value, so the more **accurate** your result.
4) There are **two types** of **error** that cause experimental uncertainty:

Random errors

1) No matter how hard you try, you **can't get rid** of random errors.
2) They can just be down to **noise**, or that you're measuring a **random process** such as nuclear radiation emission.
3) You get random error in **any** measurement. If you measured the length of a wire 20 times, the chances are you'd get a **slightly different** value each time, e.g. due to your head being in a slightly different position when reading the scale.
4) It could be that you just can't keep controlled variables **exactly** the same throughout the experiment.
5) Or it could just be the wind was blowing in the wrong direction at the time...

Systematic errors

1) You get systematic errors not because you've made a mistake in a measurement — but because of the **apparatus** you're using, or your experimental method. E.g. using an inaccurate clock.
2) The problem is often that you **don't know they're there**. You've got to spot them first to have any chance of correcting for them.
3) Systematic errors usually **shift** all of your results to be too high or too low by the **same amount**. They're annoying, but there are things you can do to reduce them if you manage to spot them...

Lorraine thought getting an uncertainty of ± 0.1 A deserved a victory dance.

You Need to Know How to Improve Measurements

There are a few different ways you can **reduce** the uncertainty in your results:

Repeating measurements — by repeating a measurement **several times** and **averaging**, you reduce the **random uncertainty** in your result. The **more** measurements you average over, the **less error** you're likely to have.

Use higher precision apparatus — the **more precisely** you can measure something, the **less random error** there is in the measurement. So if you use more precise equipment — e.g. swapping a millimetre ruler for a micrometer to measure the diameter of a wire — you can instantly cut down the **random error** in your experiment.

Calibration — you can calibrate your apparatus by measuring a **known value**. If there's a **difference** between the **measured** and **known** value, you can use this to **correct** the inaccuracy of the apparatus, and so reduce your **systematic error**.

You can Calculate the Percentage Uncertainty in a Measurement

1) You might get asked to work out the percentage uncertainty in a measurement.
2) It's just working out a percentage, so nothing too tricky. It's just that sometimes you can get **the fear** as soon as you see the word uncertainty... but just keep your cool and you can pick up some easy marks.

Example

Tom finds the resistance of a filament lamp to be **5.0 ± 0.4 Ω**.

The percentage uncertainty in the resistance measured $= \frac{0.4}{5.0} \times 100 = \mathbf{8\%}$

Error Analysis

You can Estimate Values by Averaging

You might be given a graph of information showing the results for many **repetitions** of the **same** experiment, and asked to estimate the true value and give an uncertainty in that value. Yuk. Here's how to go about it:

1) Estimate the true value by **averaging** the results you've been given. (Make sure you state whatever average it is you take, otherwise you might not get the mark.)
2) To get the uncertainty, you just need to look how far away from your average value the maximum and minimum values in the graph you've been given are.

Example — Estimating the resistance of a component

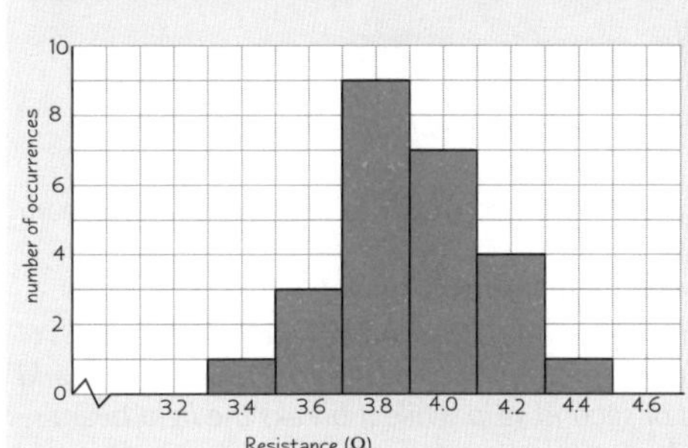

A class measure the resistance of a component and record their results on the bar chart shown. Estimate the resistance of the component, giving a suitable range of uncertainty in your answer.

There were 25 measurements, so taking the **mean**:

$$\frac{(3.4 + (3.6 \times 3) + (3.8 \times 9) + (4.0 \times 7) + (4.2 \times 4) + 4.4)}{25} = \frac{97.6}{25} = 3.90 \text{ (3 s.f.)}$$

The maximum value found was 4.4 Ω, the minimum value was 3.4. Both values are both about 0.5 Ω from the average value, so the answer is **3.9 ± 0.5 Ω**.

Error Bars to Show Uncertainty on a Graph

1) Most of the time in science, you work out the uncertainty in your **final result** using the uncertainty in **each measurement** you make.
2) When you're plotting a graph, you show the uncertainty in a value by using **error bars** to show the range the point is likely to lie in.
3) You probably won't get asked to **plot** any error bars (phew...) — but you might need to **read off** a graph that has them.

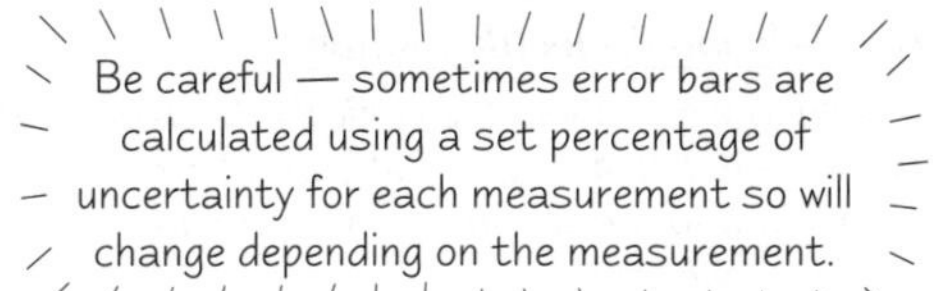

Be careful — sometimes error bars are calculated using a set percentage of uncertainty for each measurement so will change depending on the measurement.

Example

Use the graph below to find the error in measuring the extension of material X.

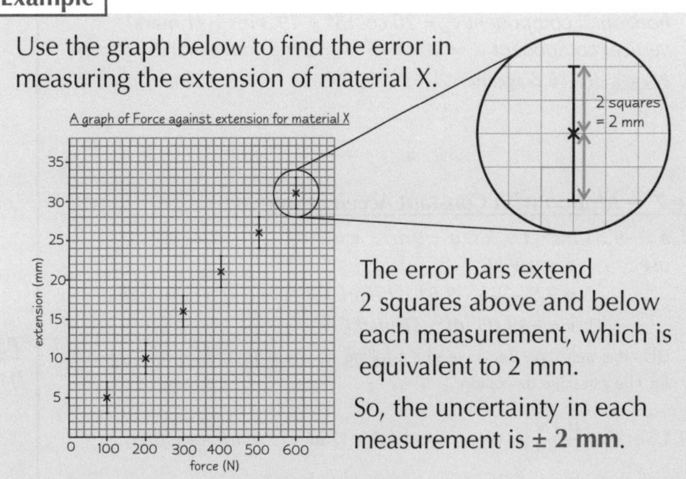

The error bars extend 2 squares above and below each measurement, which is equivalent to 2 mm.

So, the uncertainty in each measurement is **± 2 mm**.

You can Estimate the Uncertainty of the Graph's Gradient

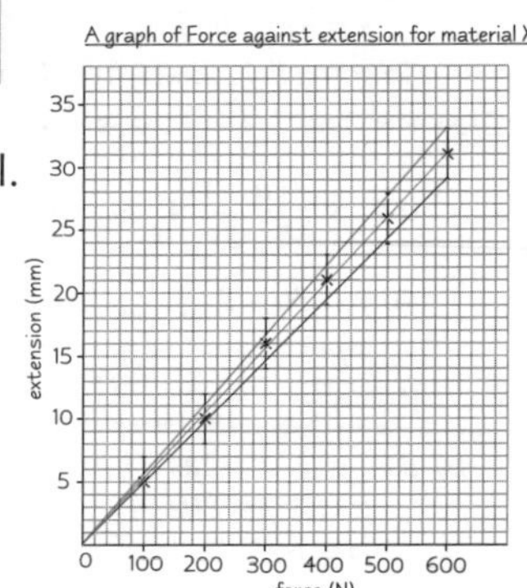

1) Normally when you draw a graph you'll want to find the gradient or intercept. E.g. for a force-extension graph, the gradient's $1/\boldsymbol{k}$, the stiffness constant of the material.
2) To find the value of $\boldsymbol{k}$, you draw a nice line of best fit on the graph and calculate your answer from that. No problem there.
3) You can then draw the **maximum** and **minimum** slopes possible for the data through **all** of the error bars. By calculating the value of the gradient (or intercept) for these slopes, you can find maximum and minimum values the true answer is likely to lie between. And that's the **uncertainty** in your answer.

Random error in your favour — collect £200...

These pages should give you a fair idea of how to deal with errors... which are rather annoyingly in everything. Even if you're lucky enough to not get tested on this sort of thing in the exam, it's really useful to know for your lab coursework.

Answers

Unit 1: Section 1 — Motion

Page 5 — Scalars and Vectors

1) Start by drawing a diagram:

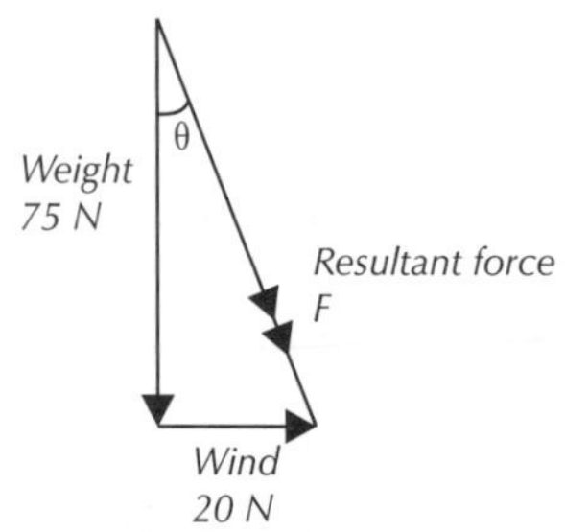

$F^2 = 20^2 + 75^2 = 6025$
So $F = 77.6$ N
$\tan\theta = 20 / 75 = 0.267$
So $\theta = 14.9°$
The resultant force on the rock is 77.6 N [1 mark] at an angle of 14.9° [1 mark] to the vertical.

Make sure you know which angle you're finding — and label it on your diagram.

2) Again, start by drawing a diagram:

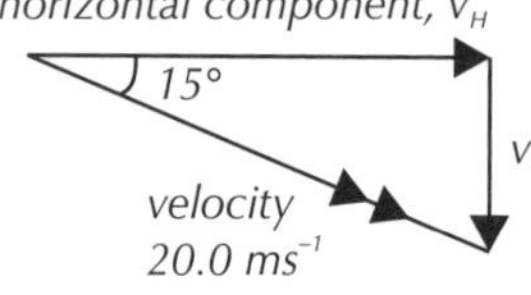

horizontal component $v_H = 20 \cos 15° = 19.3\ ms^{-1}$ [1 mark]
vertical component $v_V = 20 \sin 15° = 5.2\ ms^{-1}$ downwards [1 mark]

Always draw a diagram.

Page 7 — Motion with Constant Acceleration

1) a) $a = -9.81\ ms^{-2}$, $t = 5$ s, $u = 0\ ms^{-1}$, $v = ?$
use: $v = u + at$
$v = 0 + 5 \times -9.81$ [1 mark for either step of working]
$v = -49.05\ ms^{-1}$ [1 mark]

NB: It's negative because she's falling downwards and we took upwards as the positive direction.

b) Use: $s = \left(\frac{u+v}{2}\right)t$ or $s = ut + \frac{1}{2}at^2$ [1 mark for either]

$s = \frac{-49.05}{2} \times 5$ $\quad$ $s = 0 + \frac{1}{2} \times -9.81 \times 5^2$

$s = -122.625$ m $\quad$ $s = -122.625$ m

So she fell 122.625 m [1 mark for answer]

2) a) $v = 0\ ms^{-1}$, $t = 3.2$ s, $s = 40$ m, $u = ?$

use: $s = \left(\frac{u+v}{2}\right)t$ [1 mark]

$40 = 3.2u \div 2$

$u = \frac{80}{3.2} = 25\ ms^{-1}$ [1 mark]

b) use: $v^2 = u^2 + 2as$ [1 mark]
$0 = 25^2 + 80a$
$-80a = 625$
$a = -7.8\ ms^{-2}$ [1 mark]

3) a) Take upstream as negative: $v = 5\ ms^{-1}$, $a = 6\ ms^{-2}$, $s = 1.2$ m, $u = ?$
use: $v^2 = u^2 + 2as$ [1 mark]
$5^2 = u^2 + 2 \times 6 \times 1.2$
$u^2 = 25 - 14.4 = 10.6$
$u = -3.26\ ms^{-1}$ [1 mark]

b) From furthest point: $u = 0\ ms^{-1}$, $a = 6\ ms^{-2}$, $v = 5\ ms^{-1}$, $s = ?$
use: $v^2 = u^2 + 2as$ [1 mark]
$5^2 = 0 + 2 \times 6 \times s$
$s = 25 \div 12 = 2.08$ m [1 mark]

Page 9 — Free Fall

1) a) The computer needs:
The time for the first strip of card to pass through the beam [1 mark]
The time for the second strip of card to pass through the beam [1 mark]
The time between these events [1 mark]

b) Average speed of first strip while it breaks the light beam = width of strip ÷ time to pass through beam [1 mark]
Average speed of second strip while it breaks the light beam = width of strip ÷ time to pass through beam [1 mark]
Acceleration = (second speed – first speed) ÷ time between light beam being broken [1 mark]

c) E.g. the device will accelerate while the beam is broken by the strips. [1 mark]

2) a) You know $s = 5$ m, $a = -g$, $v = 0$
You need to find u, so use $v^2 = u^2 + 2as$
$0 = u^2 - 2 \times 9.81 \times 5$ [1 mark for either line of working]
$u^2 = 98.1$, so $u = 9.9\ ms^{-1}$ [1 mark]

b) You know $a = -g$, $v = 0$ at highest pt, $u = 9.9\ ms^{-1}$ from a)
You need to find t, so use $v = u + at$
$0 = 9.9 - 9.81t$ [1 mark for either line of working]
$t = 9.9/9.81 = 1.0$ s [1 mark]

c) Her velocity as she lands back on the trampoline will be $-9.9\ ms^{-1}$ (same magnitude, opposite direction)
[2 marks — 1 for correct number, 1 for correct sign]

Page 11 — Free Fall and Projectile Motion

1) a) You only need to worry about the vertical motion of the stone.
$u = 0\ ms^{-1}$, $s = -560$ m, $a = -g = -9.81\ ms^{-2}$, $t = ?$
You need to find t, so use: $s = ut + \frac{1}{2}at^2$ [1 mark]
$-560 = 0 + \frac{1}{2} \times -9.81 \times t^2$

$t = \sqrt{\frac{2 \times (-560)}{-9.81}} = 10.7\text{s (1 d.p.)} = 11\text{ s (to the nearest second)}$

[1 mark]

b) You know that in the horizontal direction:
$v = 20$ m/s, $t = 10.7$ s, $a = 0$, $s = ?$

So use velocity $= \frac{\text{distance}}{\text{time}}$, $v = \frac{s}{t}$ [1 mark]

$s = v \times t = 20 \times 10.7 = 214$ m (to the nearest metre) [1 mark]

Answers

2) *You know that for the arrow's vertical motion (taking upwards as the positive direction):*
$\mathbf{a} = -9.81$ ms^{-2}, $\mathbf{u} = 30$ ms^{-1} and the arrow will be at its highest point just before it starts falling back towards the ground, so $\mathbf{v} = 0$ m/s.
$\mathbf{s}$ = the distance travelled from the arrow's firing point
So use $\mathbf{v}^2 = \mathbf{u}^2 + 2\mathbf{as}$ [1 mark]
$0 = 30^2 + 2 \times -9.81 \times \mathbf{s}$

$$900 = 2 \times 9.81\mathbf{s}$$

$$\mathbf{s} = \frac{900}{2 \times 9.81} = 45.9 \text{ m} \quad [1 \text{ mark}]$$

So the maximum distance reached from the ground = 45.9 + 1 = 47 m (to the nearest metre). [1 mark]

Page 13 — Displacement-Time Graphs

1) *Split graph into four sections:*

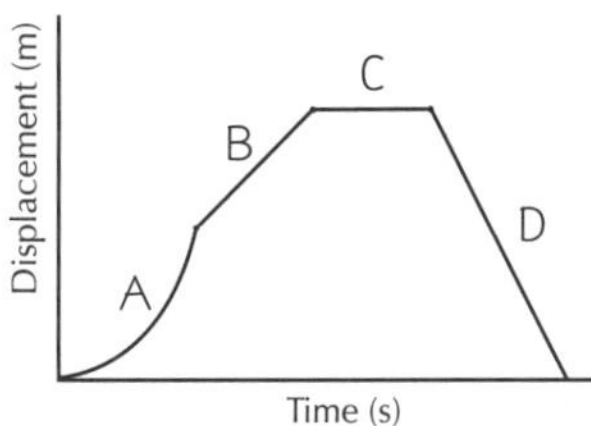

A: acceleration [1 mark]
B: constant velocity [1 mark]
C: stationary [1 mark]
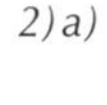
D: constant velocity in opposite direction to A and B [1 mark]

2) a)

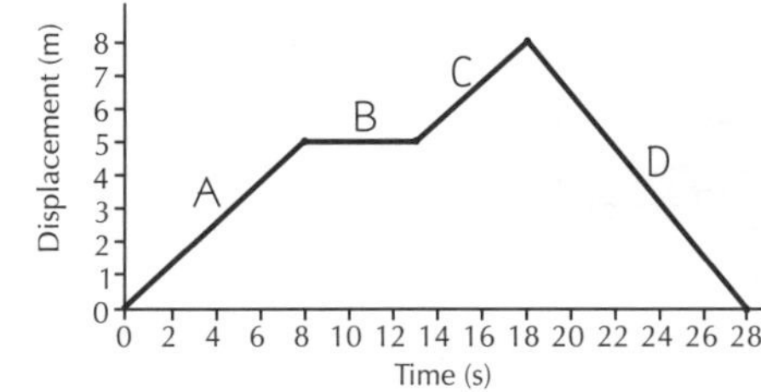

[4 marks — 1 mark for each section correctly drawn]

b) *At A:* $\mathbf{v} = \frac{\text{displacement}}{\text{time}} = \frac{5}{8} = 0.625$ ms^{-1}

At B: $\mathbf{v} = 0$

At C: $\mathbf{v} = \frac{\text{displacement}}{\text{time}} = \frac{3}{5} = 0.6$ ms^{-1}

At D: $\mathbf{v} = \frac{\text{displacement}}{\text{time}} = \frac{-8}{10} = -0.8$ ms^{-1}

[2 marks for all correct or just 1 mark for 2 or 3 correct]

Page 15 — Velocity-Time Graphs

1) a)

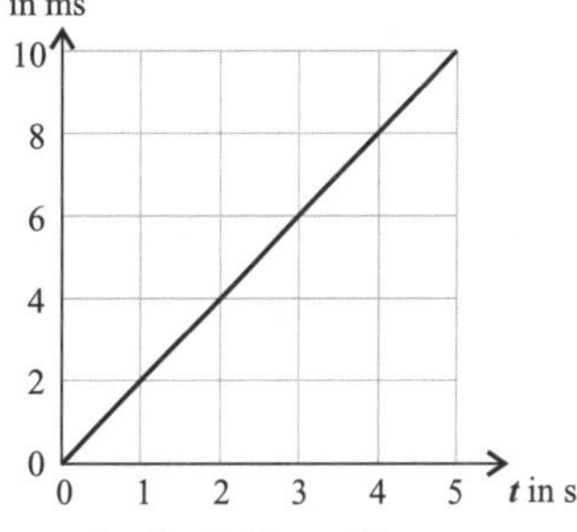

[2 marks]

b) *use $\mathbf{s} = \mathbf{ut} + \frac{1}{2}\mathbf{at}^2$ [1 mark]*
$\mathbf{t} = 1$, $\mathbf{s} = 1$
$\mathbf{t} = 2$, $\mathbf{s} = 4$
$\mathbf{t} = 3$, $\mathbf{s} = 9$
$\mathbf{t} = 4$, $\mathbf{s} = 16$
$\mathbf{t} = 5$, $\mathbf{s} = 25$
[2 marks for all correct or 1 mark for at least 3 pairs of values right]

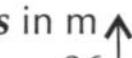
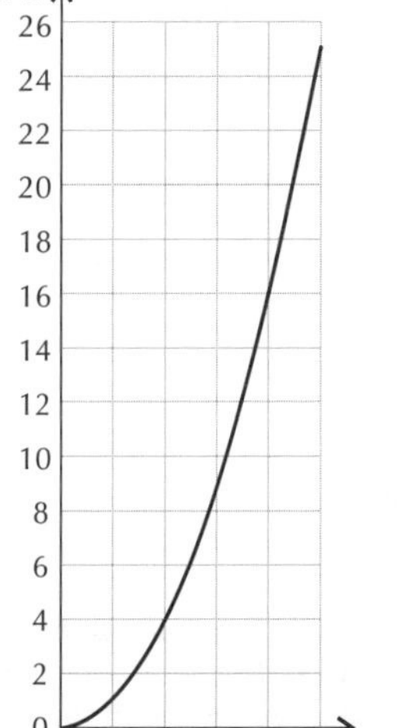

[1 mark for correctly labelled axes, 1 mark for correct curve]

c) *E.g. another way to calculate displacement is to find the area under the velocity-time graph. [1 mark]*
E.g. total displacement = ½ × 5 × 10 = 25 m [1 mark]

Answers

Unit 1: Section 2 — Forces in Action

Page 17 — Newton's Laws of Motion

1) a) b)

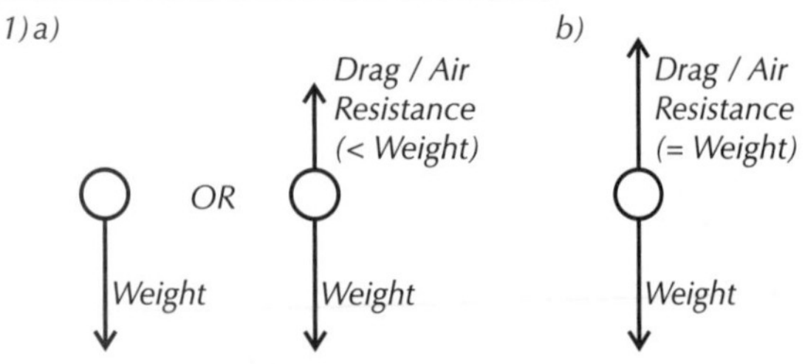

[1 mark for each diagram]

2) a) *Force perpendicular to river flow = 500 – 100 = 400 N [1 mark]*
Force parallel to river flow = 300 N
Magnitude of resultant force = $\sqrt{400^2 + 300^2}$ *= 500 N [1 mark]*

b) $\mathbf{a} = \mathbf{F}/\mathbf{m}$ (from $\mathbf{F} = \mathbf{ma}$) *[1 mark]*
= 500/250 = 2 ms^{-2} *[1 mark]*

3) a) *The resultant force acting on it [1 mark] and its mass. [1 mark]*

b) *E.g. Michael is able to exert a greater force than Tom. Michael is lighter than Tom. [1 mark each for 2 sensible points]*

c) *The only force acting on each of them is their weight =* **mg** *[1 mark]. Since* **F** = **ma**, *this gives* **ma** = **m**g, *or* **a** = g *[1 mark]. Their acceleration doesn't depend on their mass — it's the same for both of them — so they reach the water at the same time. [1 mark]*

Page 19 — Drag and Terminal Velocity

1) a) *The velocity increases at a steady rate, which means the acceleration is constant. [1 mark]*
Constant acceleration means there must be no atmospheric resistance (atmospheric resistance would increase with velocity, leading to a decrease in acceleration). So there must be no atmosphere. [1 mark]

b)

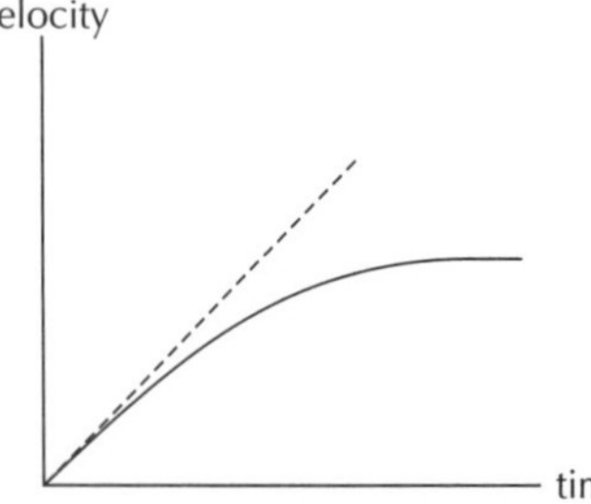

[1 mark for a smooth curve that levels out, 1 mark for correct position relative to existing line]

Your graph must be a smooth curve which levels out. It must NOT go down at the end.

c) *(The graph becomes less steep)*
because the acceleration is decreasing [1 mark]
because air resistance increases with speed [1 mark]
(The graph levels out)
because air resistance has become equal to weight [1 mark]

If the question says 'explain', you won't get marks for just describing what the graph shows — you have to say <u>why</u> it is that shape.

Page 21 — Mass, Weight and Centre of Gravity

1) a) *Density is a measure of 'compactness' of a material — its mass per unit volume. [1 mark]*

b) $\rho = \frac{\mathbf{m}}{\mathbf{V}}$ *[1 mark]*
V of cylinder = $\pi \mathbf{r}^2\mathbf{h} = \pi \times 4^2 \times 6 = 301.6\ cm^3$ *[1 mark]*
$\rho = 820 \div 301.6 = 2.72\ g\,cm^{-3}$ *[1 mark]*

c) $\mathbf{V} = 5 \times 5 \times 5 = 125\ cm^3$
$\mathbf{m} = \rho \times \mathbf{V} = 2.7 \times 125 = 340\ g$ *[1 mark]*

2) *Experiment:*
Hang the object freely from a point. Hang a plumb bob from the same point, and use it to draw a vertical line down the object. [1 mark]
Repeat for a different point and find the point of intersection. [1 mark]
The centre of gravity is halfway through the thickness of the object (by symmetry) at the point of intersection. [1 mark]
Identifying and reducing error, e.g.:
Source: the object and/or plumb line might move slightly while you're drawing the vertical line [1 mark]
Reduced by: hang the object from a third point to confirm the position of the point of intersection [1 mark]

Page 23 — Forces and Equilibrium

1)

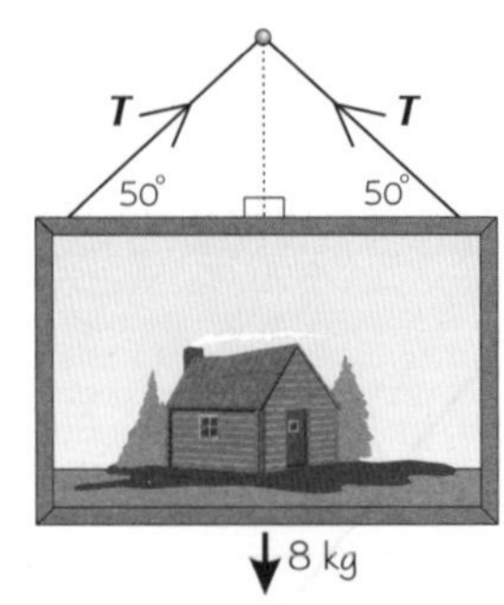

Weight = vertical component of tension × 2
$8 \times 9.81 = 2\mathbf{T}\sin 50°$ *[1 mark]*
$78.48 = 0.766 \times 2\mathbf{T}$
$102.45 = 2\mathbf{T}$
$\mathbf{T} = 51.2\ N$ *[1 mark]*

2)

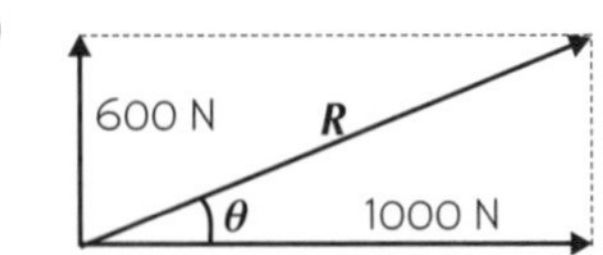

By Pythagoras:
$\mathbf{R} = \sqrt{1000^2 + 600^2} = 1166\ N$ *[1 mark]*

$\tan\theta = \frac{600}{1000}$, so $\theta = \tan^{-1} 0.6 = 31.0°$ *[1 mark]*

Answers

Page 25 — Moments and Torques

1) *Torque = Force × distance [1 mark]*
$60 = 0.4\mathbf{F}$, so $\mathbf{F} = 150$ N [1 mark]

2)

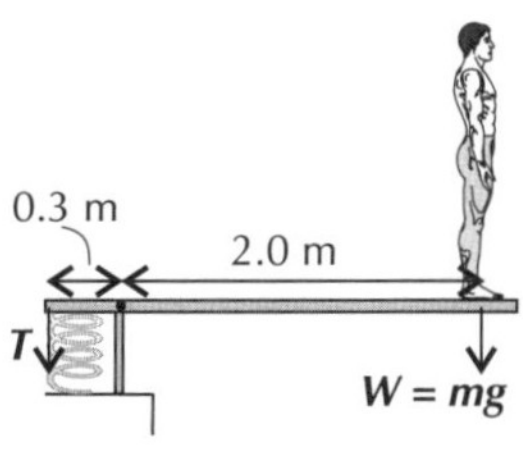

clockwise moment = anticlockwise moment
$\mathbf{W} \times 2.0 = \mathbf{T} \times 0.3$ [1 mark for either line of working]
$60 \times 9.81 \times 2.0 = \mathbf{T} \times 0.3$
$\mathbf{T} = 3924$ N [1 mark]

The tension in the spring is equal and opposite to the force exerted by the diver on the spring.

Page 27 — Car Safety

1) a) *reaction time is 0.5 s, speed is 20 ms^{-1}*
$\mathbf{s} = \mathbf{vt}$ [1 mark]
$= 20 \times 0.5 = 10$ m [1 mark]

b) *Use $\mathbf{F} = \mathbf{ma}$ to get $\mathbf{a}$: $\mathbf{a} = -10\,000/850 = -11.76\ ms^{-2}$ [1 mark]*

Use $\mathbf{v}^2 = \mathbf{u}^2 + 2\mathbf{as}$, and rearrange to get $\mathbf{s} = \dfrac{\mathbf{v}^2 - \mathbf{u}^2}{2\mathbf{a}}$

Put in the values: $\mathbf{s} = (0 - 400) \div (2 \times -11.76)$ [1 mark]
$= 17$ m [1 mark]

Remember that a force against the direction of motion is negative.

c) *Total stopping distance = 10 + 17 = 27 m*
She stops 3 m before the cow. [1 mark]

2) a) *Car: use $\mathbf{v} = \mathbf{u} + \mathbf{at}$ to get acceleration:*
$\mathbf{a} = (0 - 20)/0.1 = -200\ ms^{-2}$ [1 mark]
Use $\mathbf{F} = \mathbf{ma}$:
$\mathbf{F} = 900 \times -200 = -180\,000$ N [1 mark]
Same for dummy:
$\mathbf{a} = 0 - 18/0.1 = -180\ ms^{-2}$ [1 mark]
$\mathbf{F} = 50 \times -180 = -9000$ N [1 mark]

b) *Crumple zones will increase the collision time for the car and dummy;*
this reduces forces on the car and dummy;
the airbag will keep the dummy in its seat;
and increase the collision time further for the dummy;
reducing the force on it.
[3 marks for any three sensible points]

Unit 1: Section 3 — Work and Energy

Page 29 — Work and Power

1) a)

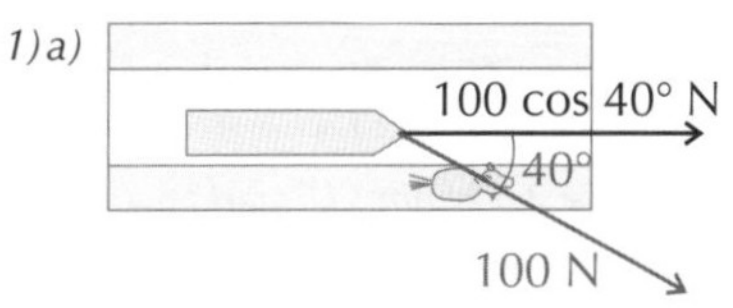

Force in direction of travel = 100 cos40° = 76.6 N [1 mark]
$\mathbf{W} = \mathbf{Fs} = 76.6 \times 1500 = 114\,900$ J [1 mark]

b) *Use $\mathbf{P} = \mathbf{Fv}$ [1 mark]*
$= 100 \cos 40° \times 0.8 = 61.3$ W [1 mark]

2) a) *Use $\mathbf{W} = \mathbf{Fs}$ [1 mark]*
$= 20 \times 9.81 \times 3 = 588.6$ J [1 mark]

Remember that 20 kg is not the force — it's the mass. So you need to multiply it by 9.81 Nkg^{-1} to get the weight.

b) *Use $\mathbf{P} = \mathbf{Fv}$ [1 mark]*
$= 20 \times 9.81 \times 0.25 = 49.05$ W [1 mark]

Page 31 — Conservation of Energy

1) a) *Use $\mathbf{E}_k = \frac{1}{2}\mathbf{mv}^2$ and $\mathbf{E}_p = \mathbf{mgh}$ [1 mark]*
$\frac{1}{2}\mathbf{mv}^2 = \mathbf{mgh}$
$\frac{1}{2}\mathbf{v}^2 = \mathbf{gh}$
$\mathbf{v}^2 = 2\mathbf{gh} = 2 \times 9.81 \times 2 = 39.24$ [1 mark]
$\mathbf{v} = 6.26\ ms^{-1}$ [1 mark]

'No friction' allows you to say that the changes in kinetic and potential energy will be the same.

b) *2 m — no friction means the kinetic energy will all change back into potential energy, so he will rise back up to the same height as he started. [1 mark]*

c) *Put in some more energy by actively 'skating'. [1 mark]*

2) a) *If there's no air resistance, $\mathbf{E}_k = \mathbf{E}_p = \mathbf{mgh}$ [1 mark]*
$\mathbf{E}_k = 0.02 \times 9.81 \times 8 = 1.57$ J [1 mark]

b) *If the ball rebounds to 6.5 m, it has gravitational potential energy:*
$\mathbf{E}_p = \mathbf{mgh} = 0.02 \times 9.81 \times 6.5 = 1.28$ J [1 mark]
So $1.57 - 1.28 = 0.29$ J is converted to other forms [1 mark]

Page 33 — Efficiency and Sankey Diagrams

1) a) *Wasted heat energy = 125 – 30 – 70 = 25 KJ [1 mark]*

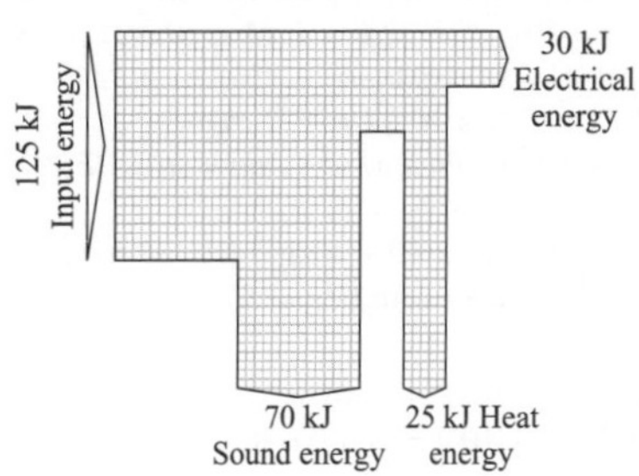

[1 mark for correctly drawn input arrow, 1 mark for correctly drawn output arrows, 1 mark for labels]

b) *Efficiency = $\dfrac{\text{useful output energy}}{\text{total input energy}} \times 100\%$*

Efficiency of first design = $\dfrac{15}{60} \times 100\% = 25\%$ [1 mark]

Efficiency of second design = $\dfrac{30}{125} \times 100\% = 24\%$ [1 mark]

The first design is 1% more efficient than the second. [1 mark]

Answers

Page 35 — Hooke's Law

1) a) *Force is proportional to extension.*
The force is 1.5 times as great, so the extension will also be 1.5 times the original value.
Extension = 1.5 × 4.0 mm = 6.0 mm [1 mark]

b) $\mathbf{F} = \mathbf{ke}$ *and so* $\mathbf{k} = \mathbf{F}/\mathbf{e}$ *[1 mark]*
$\mathbf{k} = 10 \div 4.0 \times 10^{-3} = 2500\ Nm^{-1}$ *or* $2.5\ Nmm^{-1}$ *[1 mark]*
There is one mark for rearranging the equation and another for getting the right numerical answer.

c) *One mark for any sensible point e.g.*
The string now stretches much further for small increases in force.
When the string is loosened it is longer than at the start. [1 mark]

2) *The rubber band does not obey Hooke's law [1 mark] because when the force is doubled from 2.5 N to 5 N, the extension increases by a factor of 2.3. [1 mark]*

Page 37 — Stress and Strain

1) a) *Area =* $\pi \mathbf{d}^2/4$ *or* $\pi \mathbf{r}^2$.
So area = $\pi \times (1 \times 10^{-3})^2/4 = 7.85 \times 10^{-7}\ m^2$ *[1 mark]*

b) *Stress = force/area =* $300/(7.85 \times 10^{-7}) = 3.82 \times 10^8\ Nm^{-2}$ *[1 mark]*

c) *Strain = extension/length =* $4 \times 10^{-3}/2.00 = 2 \times 10^{-3}$ *[1 mark]*

2) a) $\mathbf{F} = \mathbf{ke}$ *and so rearranging* $\mathbf{k} = \mathbf{F}/\mathbf{e}$ *[1 mark]*
So $\mathbf{k} = 50/(3.0 \times 10^{-3}) = 1.67 \times 10^4\ Nm^{-1}$ *[1 mark]*

b) *Elastic strain energy =* $\frac{1}{2}\mathbf{Fe}$
Giving the elastic strain energy as
$\frac{1}{2} \times 50 \times 3 \times 10^{-3} = 7.5 \times 10^{-2}\ J$ *[1 mark]*

3) *Elastic strain energy,*
$\mathbf{E} = \frac{1}{2}\mathbf{ke}^2 = \frac{1}{2} \times 40.8 \times 0.05^2 = 0.051\ J$ *[1 mark]*
To find maximum speed, assume all this energy is converted to kinetic energy in the ball. $\mathbf{E}_{kinetic} = \mathbf{E}$ *[1 mark]*
$\mathbf{E} = \frac{1}{2}\mathbf{mv}^2$*, so rearranging,* $\mathbf{v}^2 = 2\mathbf{E}/\mathbf{m}$ *[1 mark]*
$\mathbf{v}^2 = (2 \times 0.051)/0.012 = 8.5$*, so* $\mathbf{v} = 2.92\ ms^{-1}$ *[1 mark]*

Page 39 — The Young Modulus

1) a) *Cross-sectional area =* $\pi \mathbf{d}^2/4$ *or* $\pi \mathbf{r}^2$.
So the cross-sectional area = $\pi \times (0.6 \times 10^{-3})^2/4 = 2.83 \times 10^{-7}\ m^2$ *[1 mark]*

b) *Stress = force/area =* $80/(2.83 \times 10^{-7}) = 2.83 \times 10^8\ Nm^{-2}$ *[1 mark]*

c) *Strain = extension/length =* $3.6 \times 10^{-3}/2.5 = 1.44 \times 10^{-3}$ *[1 mark]*

d) *The Young modulus for steel = stress/strain*
$= 2.83 \times 10^8/(1.44 \times 10^{-3}) = 2.0 \times 10^{11}\ Nm^{-2}$ *[1 mark]*

2) a) *The Young modulus,* $\mathbf{E}$ *= stress/strain and so strain = stress/*$\mathbf{E}$ *[1 mark]*
Strain on copper = $2.6 \times 10^8/1.3 \times 10^{11} = 2 \times 10^{-3}$ *[1 mark]*
There's one mark for rearranging the equation and another for using it.

b) *Stress = force/area and so area = force/stress*
Area of the wire = $100/(2.6 \times 10^8) = 3.85 \times 10^{-7}\ m^2$ *[1 mark]*

c) *Strain energy per unit volume =* $\frac{1}{2}$ *× stress × strain*
$= \frac{1}{2} \times 2.6 \times 10^8 \times 2 \times 10^{-3} = 2.6 \times 10^5\ Jm^{-3}$ *[1 mark]*
Give the mark if answer is consistent with the value calculated for strain in part a).

Page 41 — Interpreting Stress-Strain Graphs

1) a) *Liable to break suddenly without deforming plastically. [1 mark]*

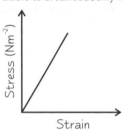

b)
Stress (Nm^{-2})
Strain

[1 mark for correctly labelled axes, 1 mark for straight line through the origin]

2) a)

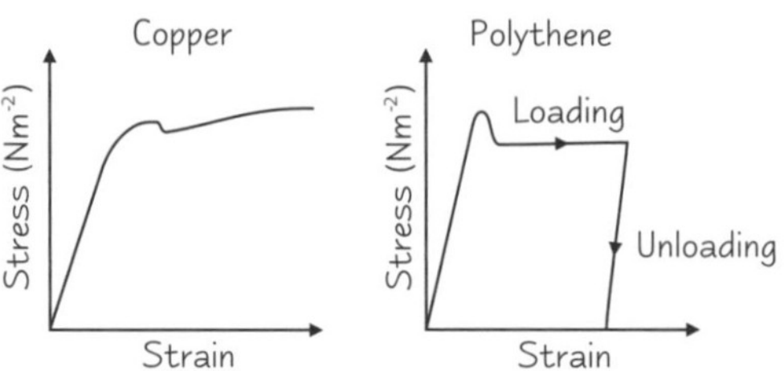

[1 mark for correctly labelled axes, 1 mark for correct shape of copper graph, 1 mark for correct shape of polythene graph (loading curve only is acceptable]

b) *The stress-strain graphs for both materials begin with a straight line through the origin, showing that both materials initially obey Hooke's law. [1 mark]*
Both materials undergo plastic deformation when a large enough stress is applied. [1 mark]

Answers

Unit 2: Section 1 — Electric Circuits, Resistance and DC Circuits

Page 43 — Charge, Current and Potential Difference

1) Time in seconds = 10 × 60 = 600 s.
Use the formula $I = Q / t$ [1 mark]
which gives you $I = 4500 / 600 = 7.5$ A [1 mark]
Write down the formula first. Don't forget the unit in your answer.

2) Rearrange the formula $I = nAvq$ and you get $v = I / nAq$ [1 mark]
which gives you

$$v = \frac{13}{(1.0 \times 10^{29}) \times (5.0 \times 10^{-6}) \times (1.6 \times 10^{-19})}$$ [1 mark]

$v = 1.63 \times 10^{-4}\ ms^{-1}$ [1 mark]

3) Work done = 0.75 × electrical energy input
so the energy input will be 90 / 0.75 = 120 J. [1 mark]
Rearrange the formula $V = W / Q$ to give $Q = W / V$ [1 mark]
so you get $Q = 120 / 12 = 10$ C. [1 mark]
The electrical energy input to a motor has to be greater than the work it does because motors are less than 100% efficient.

Page 45 — Resistance and Resistivity

1) Area = $\pi(d/2)^2$ and $d = 1.0 \times 10^{-3}$ m
so Area = $\pi \times (0.5 \times 10^{-3})^2 = 7.85 \times 10^{-7} m^2$ [1 mark]

$$R = \frac{\rho l}{A} = \frac{2.8 \times 10^{-8} \times 4}{7.85 \times 10^{-7}} = 0.14\ \Omega$$

[1 mark for equation or working, 1 mark for answer with unit.]

2) a) $R = V / I$ [1 mark]

$$= \frac{2}{2.67 \times 10^{-3}} = 749\ \Omega$$ [1 mark]

b) Two further resistance calculations give 750 Ω for each answer [1 mark]
There is no significant change in resistance for different potential differences [1 mark]
Component is an ohmic conductor because its resistance is constant for different potential differences. [1 mark]

Page 47 — I/V Characteristics

1) a)

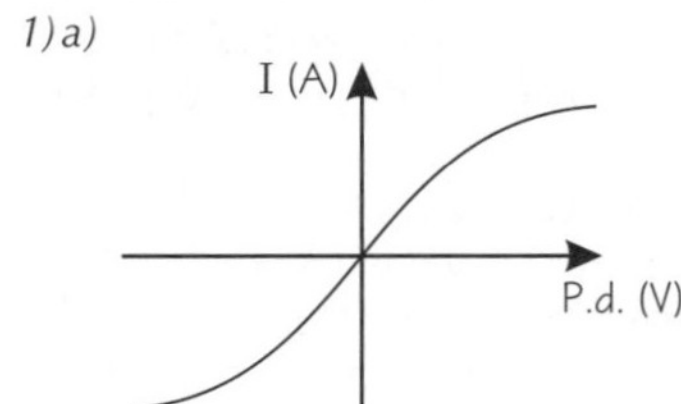

[1 mark]

b) Resistance increases as the temperature increases [1 mark]

c) Increase in temperature makes metal ions vibrate more [1 mark]
Increased collisions with ions impedes electrons [1 mark]

Page 49 — Electrical Energy and Power

1) a) $I = P/V$ [1 mark] = 920/230 = 4 A [1 mark]

b) $I = V/R$ [1 mark] = 230/190 = 1.21 A [1 mark]

c) $P_{motor} = VI = 230 \times 1.21 = 278$ W [1 mark]
Total power = motor power + heater power
= 278 + 920 = 1198 W, which is approx. 1.2 kW [1 mark]

2) a) Energy supplied = $VIt = 12 \times 48 \times 2 = 1152$ J [1 mark]

b) Energy lost = I^2Rt [1 mark] = $48^2 \times 0.01 \times 2 = 46$ J [1 mark]

Page 51 — Domestic Energy and Fuses

1) a) Energy = power × time
i) Energy = 1800 × (15 × 60) = 1 620 000 J (= 1.62 MJ) [1 mark]
ii) Energy = 1.8 × (15 ÷ 60) = 0.45 kWh [1 mark]

b) Cost = number of units × price per unit = 0.45 × 14.6 = 6.57p [1 mark]

2) a) P = VI. Rearranging, I = P/V = 1500 ÷ 230 = 6.52 A [1 mark]
A 13 A fuse should be used. [1 mark]

b) E = Pt = 1.5 × 2.25 = 3.375 kWh [1 mark]
Cost = 3.375 × 9.8 = 33.075p [1 mark]

c) P = VI = 230 × (6.5 × 10^{-3}) = 1.495 ≈ 1.5 W [1 mark]

d) E = Pt = 0.0015 × 10 = 0.015 kWh [1 mark]
Cost = 0.015 × 9.8 = 0.147p [1 mark]

Page 53 — E.m.f. and Internal Resistance

1) a) Total resistance = $R + r = 4 + 0.8 = 4.8\ \Omega$ [1 mark]
I = e.m.f./total resistance = 24/4.8 = 5 A [1 mark]

b) $V = \varepsilon - Ir = 24 - 5 \times 0.8 = 20$ V [1 mark]

2) a) $\varepsilon = I(R + r)$, so $r = \varepsilon/I - R$ [1 mark]
$r = 500/(50 \times 10^{-3}) - 10 = 9990\ \Omega$ [1 mark]

b) This is a very high internal resistance [1 mark]
So only small currents can be drawn, reducing the risk to the user [1 mark]

Page 55 — Conservation of Energy & Charge in Circuits

1) a) Resistance of parallel resistors:
$1/R_{parallel} = 1/6 + 1/3 = 1/2$
$R_{parallel} = 2\ \Omega$ [1 mark]
Total resistance:
$R_{total} = 4 + R_{parallel} = 4 + 2 = 6\ \Omega$ [1 mark]

b) $V = IR$, so rearranging $I_3 = V / R_{total}$ [1 mark]
$I_3 = 12 / 6 = 2$ A [1 mark]

c) $V = IR = 2 \times 4 = 8$ V [1 mark]

d) E.m.f. = sum of p.d.s in circuit, so $12 = 8 + V_{parallel}$
$V_{parallel} = 12 - 8 = 4$ V [1 mark]

e) Current = p.d. / resistance
$I_1 = 4 / 3 = 1.33$ A [1 mark]
$I_2 = 4 / 6 = 0.67$ A [1 mark]

Answers

Page 57 — The Potential Divider

1) Parallel circuit, so p.d. across both sets of resistors is 12 V.

i) V_{AB} = ½ × 12 = 6 V [1 mark]

ii) V_{AC} = 2/3 × 12 = 8 V [1 mark]

iii) $V_{BC} = V_{AC} - V_{AB}$ = 8 − 6 = 2 V [1 mark]

2) a) V_{AB} = 50/80 × 12 = 7.5 V [1 mark]
(ignore the 10 Ω — no current flows that way)

b) Total resistance of the parallel circuit:
$1/R_T$ = 1/50 + 1/(10 + 40) = 1/25
R_T = 25Ω [1 mark]
p.d. over the whole parallel arrangement = 25/55 × 12 = 5.45 V [1 mark]
p.d. across AB = 40/50 × 5.45 = 4.36 V [1 mark]
*current through 40 Ω resistor = **V/R** = 4.36/40 = 0.11 A [1 mark]*

Unit 2: Section 2 — Waves

Page 59 — The Nature of Waves

1) a) Use $v = \lambda f$ and $f = 1 / T$
So $v = \lambda / T$, giving $\lambda = vT$ [1 mark]
$\lambda = 3\ ms^{-1} \times 6\ s = 18\ m$ [1 mark]
The vertical movement of the buoy is irrelevant to this part of the question.

b) The trough to peak distance is twice the amplitude, so the amplitude is 0.6 m [1 mark]

Page 61 — Longitudinal and Transverse Waves

1) a) [This question could equally well be answered using diagrams.]
For ordinary light, the EM field vibrates in all planes at right angles to the direction of travel. [1 mark]
Iceland spar acts as a polariser. [1 mark]
When light is shone through the first disc, it only allows through vibrations in one particular plane, so emerges less bright. [1 mark]
As the two crystals are rotated relative to each other there comes a point when the allowed planes are at right angles to each other. [1 mark]
So all the light is blocked. [1 mark]
Try to remember to say that for light and other EM waves it's the electric and magnetic <u>fields</u> that vibrate.

b) Using Malus' law, $I = I_0 \cos^2 \sigma$ [1 mark]. I is half the size of I_0, so $I/I_0 = 0.5$ [1 mark]. Which means $\cos^2\sigma = 0.5$, so $\cos \sigma = 0.707...$ and $\sigma = 45°$ [1 mark]

2) E.g. Polarising filters are used in photography to remove unwanted reflections [1 mark]. Light is partially polarised when it reflects so putting a polarising filter over the lens at 90 degrees to the plane of polarisation will block most of the reflected light. [1 mark].

Page 63 — The Electromagnetic Spectrum

1) At the same speed. [1 mark]
*Both are electromagnetic waves and hence travel at **c** in a vacuum. [1 mark]*

2) a) Medical X-rays [1 mark] rely on the fact that X-rays penetrate the body well but are blocked by bone. [1 mark]
OR
Security scanners at airports [1 mark] rely on the fact that X-rays penetrate suitcases and clothes but are blocked by metal e.g. of a weapon. [1 mark]

b) The main difference between gamma rays and X-rays is that gamma rays arise from nuclear decay [1 mark] but X-rays are generated when metals are bombarded with electrons. [1 mark]

3) Any of: unshielded microwaves, excess heat, damage to eyes from too bright light, sunburn or skin cancer from UV, cancer or eye damage due to ionisation by X-rays or gamma rays.
[1 mark for the type of EM wave, 1 mark for the danger to health]

Page 65 — Superposition and Coherence

1) a) The frequencies and wavelengths of the two sources must be equal [1 mark] and the phase difference must be constant. [1 mark]

b) Interference will only be noticeable if the amplitudes of the two waves are approximately equal. [1 mark]

2) a) 180° (or 180° + 360n°). [1 mark]

b) The displacements and velocities of the two points are equal in size [1 mark] but in opposite directions. [1 mark]

Answers

Page 67 — Standing (Stationary) Waves

1) a)

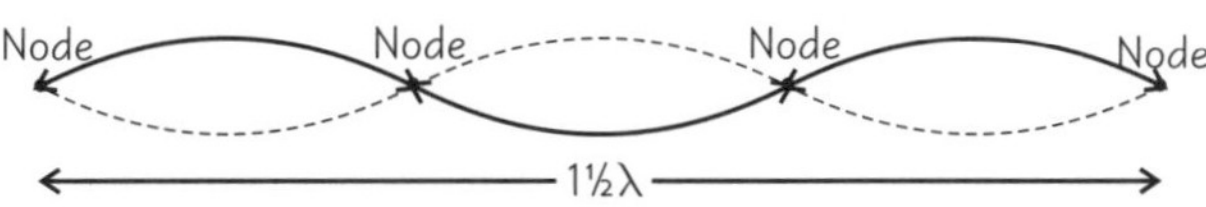

[1 mark for the correct shape, 1 mark for labelling the length]

b) For a string vibrating at three times the fundamental frequency, length = $3\lambda / 2$

$1.2 \text{ m} = 3\lambda / 2$

$\lambda = 0.8 \text{ m}$ [1 mark]

c) When the string forms a standing wave, its amplitude varies from a maximum at the antinodes to zero at the nodes. [1 mark] In a progressive wave all the points have the same amplitude. [1 mark]

Page 69 — Diffraction

1) When a wavefront meets an obstacle, the waves will diffract round the corners of the obstacle. When the obstacle is much bigger than the wavelength, little diffraction occurs. In this case, the mountain is much bigger than the wavelength of short-wave radio. So the "shadow" where you cannot pick up short wave is very long. [1 mark]

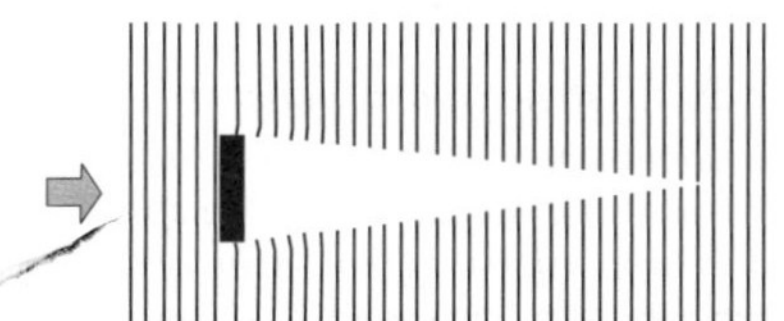

[1 mark]

When the obstacle is comparable in size to the wavelength, as it is for the long-wave radio waves, more diffraction occurs. The wavefront re-forms after a shorter distance, leaving a shorter "shadow". [1 mark]

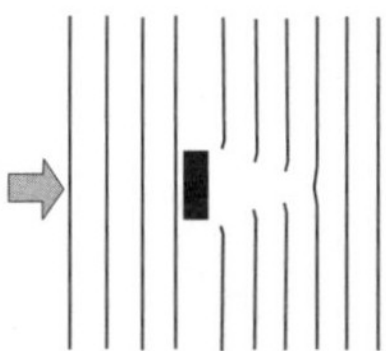

[1 mark]

Page 71 — Two-Source Interference

1) a)

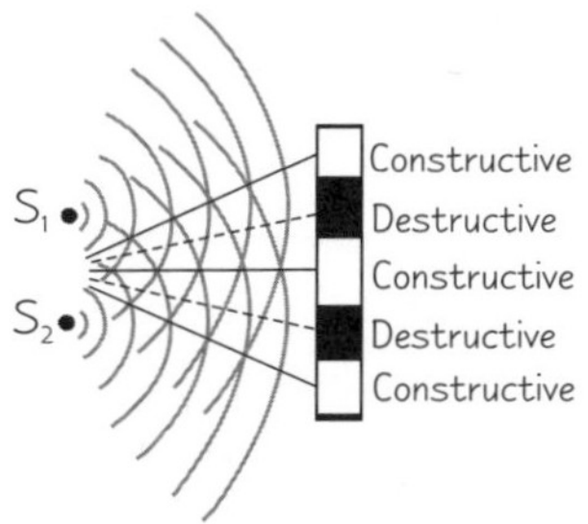

[1 mark for correct constructive interference patterns, 1 mark for correct destructive interference patterns]

b) Light waves from separate sources are not coherent, as light is emitted in random bursts of energy. To get coherent light the two sets of waves must emerge from one source. [1 mark] A laser is used because it emits coherent light that is all of one wavelength. [1 mark]

2) a) $\lambda = \mathbf{v} / \mathbf{f} = 330 / 1320 = 0.25 \text{ m}$. [1 mark]

b) Separation $= \mathbf{X} = \lambda \mathbf{D} / \mathbf{d}$ [1 mark]

$= 0.25 \text{ m} \times 7 \text{ m} / 1.5 \text{ m} = 1.17 \text{ m}$. [1 mark]

Page 73 — Diffraction Gratings

1) a) Use $\sin \theta = \mathbf{n}\lambda / \mathbf{d}$

For the first order, $\mathbf{n} = 1$

So, $\sin \theta = \lambda / \mathbf{d}$ [1 mark]

No need to actually work out d. The number of lines per metre is 1 / d. So you can simply multiply the wavelength by that.

$\sin \theta = 600 \times 10^{-9} \times 4 \times 10^{5} = 0.24$

$\theta = 13.9°$ [1 mark]

For the second order, $\mathbf{n} = 2$ and $\sin \theta = 2\lambda / \mathbf{d}$. [1 mark]

You already have a value for λ / d. Just double it to get sin θ for the second order.

$\sin \theta = 0.48$

$\theta = 28.7°$ [1 mark]

b) No. Putting $\mathbf{n} = 5$ into the equation gives a value of sin θ of 1.2, which is impossible. [1 mark]

2) $\sin \theta = \mathbf{n}\lambda / \mathbf{d}$, so for the 1st order maximum, $\sin \theta = \lambda / \mathbf{d}$ [1 mark]

$\sin 14.2° = \lambda \times 3.7 \times 10^{5}$

$\lambda = 663 \text{ nm}$ (or 6.63×10^{-7} m) [1 mark].

Answers

Unit 2: Section 3 — Quantum Phenomena

Page 75 — Light — Wave or Particle

1)a) At threshold voltage: $E_{kinetic}$ of an electron = E_{photon} emitted [1 mark]
So $E_{photon} = e \times V = 1.6\times10^{-19} \times 1.7 = 2.72 \times 10^{-19}$ J [1 mark]

b) $E = \frac{hc}{\lambda}$ [1 mark], so $h = \frac{E\lambda}{c}$
$\lambda = 7.0\times10^{-7}$ m, $c = 3.0\times10^{8}$ ms^{-1},
So, $h = \frac{2.72\times10^{-19}\times7.0\times10^{-7}}{3.0\times10^{8}} = 6.3\ \ 10^{-34}$ Js [1 mark]

Page 77 — The Photoelectric Effect

1) $\phi = 2.9\text{ eV} = 2.9\times(1.6\times10^{-19})\text{ J} = 4.64\times10^{-19}\text{ J}$ [1 mark]
$f = \frac{\phi}{h} = \frac{4.64\times10^{-19}}{6.6\times10^{-34}} = 7.0\times10^{14}$ Hz (to 2 s.f.) [1 mark]

2)a) $E = hf$ [1 mark]
$= (6.6\times10^{-34})\times(2.0\times10^{15}) = 1.32\times10^{-18}$ J [1 mark]
$1.32\times10^{-18}\text{ J} = \frac{1.32\times10^{-18}}{1.6\times10^{-19}}\text{ eV} = 8.25\text{ eV}$ [1 mark]

b) $E_{photon} = E_{max\ kinetic} + \phi$ [1 mark]
$E_{max\ kinetic} = E_{photon} - \phi = 8.25 - 4.7 = 3.55\text{ eV}$ (or 5.68×10^{-19} J)
[1 mark]

3) An electron needs to gain a certain amount of energy (the work function energy) before it can leave the surface of the metal
[1 mark]
If the energy carried by each photon is less than this work function energy, no electrons will be emitted [1 mark].

Page 79 — Energy Levels and Photon Emission

1)a) i) $E = V = 12.1$ eV [1 mark]
ii) $E = V \times 1.6 \times 10^{-19} = 12.1 \times 1.6 \times 10^{-19} = 1.9 \times 10^{-18}$ J [1 mark]

b) i) Excitation occurs when an electron moves from a lower energy to a higher energy level by absorbing energy. [1 mark]
ii) –13.6 + 12.1 = – 1.5 eV. This corresponds to $n = 3$. [1 mark]
iii) $n = 3 \rightarrow n = 2$: 3.4 – 1.5 = 1.9 eV [1 mark]
$n = 2 \rightarrow n = 1$: 13.6 – 3.4 = 10.2 eV [1 mark]
$n = 3 \rightarrow n = 1$: 13.6 – 1.5 = 12.1 eV [1 mark]

Page 81 — Wave-Particle Duality

1)a) Electromagnetic radiation can show characteristics of both a particle and a wave. [1 mark]

b) i) $E_{photon} = \frac{hc}{\lambda} = \frac{6.63\times10^{-34}\times3.00\times10^{8}}{590\times10^{-9}}$ [1 mark]
$= 3.37 \times 10^{-19}$ J [1 mark]
$E\text{ (in eV)} = \frac{E\text{ (in J)}}{1.6\times10^{-19}} = \frac{3.37\times10^{-19}}{1.6\times10^{-19}} = 2.11\text{ eV}$ [1 mark]

ii) $\lambda = \frac{h}{mv}$
$\therefore v = \frac{h}{m\lambda} = \frac{6.63\times10^{-34}}{9.1\times10^{-31}\times590\times10^{-9}} = 1230\text{ ms}^{-1}$
[2 marks, otherwise 1 mark for some correct working]

2)a) $\lambda = \frac{h}{mv} = \frac{6.63\times10^{-34}}{9.1\times10^{-31}\times3.5\times10^{6}} = 2.08\times10^{-10}$ m
[2 marks, otherwise 1 mark for some correct working]

b) Either $v = \frac{h}{m\lambda}$ with $m_{proton} = 1840 \times m_{electron}$
or momentum of protons = momentum of electrons
$1840\times \cancel{m_e} \times v_p = \cancel{m_e} \times 3.5\times10^{6}$
$v_p = 1900\text{ ms}^{-1}$
[2 marks, otherwise 1 mark for some correct working]

c) The two have the same kinetic energy if the voltages are the same. The proton has a larger mass, so it will have a smaller speed.
[1 mark] Kinetic energy is proportional to the square of the speed, while momentum is proportional to the speed, so they will have different momenta. [1 mark]
Wavelength depends on the momentum, so the wavelengths are different. [1 mark]
This is a really hard question. If you didn't get it right, make sure you understand the answer fully. Do the algebra if it helps.

3)a) $E_k = 6 \times 10^{3}$ eV [1 mark]
$= 6000 \times 1.6 \times 10^{-19} = 9.6 \times 10^{-16}$ J [1 mark]

b) $E_k = \frac{1}{2}mv^2$
$9.6\times10^{-16} = \frac{1}{2}\times9.1\times10^{-31}\times v^2$
$v = \sqrt{\frac{2\times9.6\times10^{-16}}{9.1\times10^{-31}}} = 4.6\times10^{7}\text{ ms}^{-1}$
[2 marks, otherwise 1 mark for some correct working]

c) $\lambda = \frac{h}{mv} = \frac{6.63\times10^{-34}}{9.1\times10^{-31}\times4.6\times10^{7}} = 1.58\times10^{-11}$ m
[2 marks, otherwise 1 mark for some correct working]

Index

A

acceleration 6–8, 12–18, 26, 27
acceleration-time graphs 18
air resistance 18, 19
airbags 26, 27
ammeter 42
amplitude 58, 61
antinodes 66, 67
Aristotle 10
atoms 35, 36

B

balanced forces 16, 19, 22
batteries 52
bias (of a diode) 47
braking distance 26
braking force 26
breaking stress 36
brightness 58, 72
brittle fracture 40
brittle materials 40

C

cancer 63
car safety 26, 27
centre of gravity 20, 21
charge 42, 43
charge carriers 42, 43, 46
chemical energy 52
codswallop 75
coherence 64, 65, 70
components (resolving) 5, 11, 22, 23
compressions (of longitudinal waves) 60
compressive forces 34, 36
conservation of charge 54, 55
conservation of energy 30–32, 52, 54, 55
constant acceleration equations 6, 7, 9, 11
constructive interference 64, 65, 70
contact friction 18
continuous spectra 79
coulomb (definition) 42
couple (force) 25
crumple zones 26
current 42–46, 48, 51–55

D

de Broglie wavelength 80, 81
destructive interference 64, 65, 70
diffraction 62, 68–73, 80, 81
diffraction gratings 72, 73
diodes 47
displacement 4, 6, 12, 13, 15, 37, 58, 64
displacement-time graphs 12, 13, 15
distance 14, 28
domestic energy 50, 51
drag (fluid friction) 18, 19
drift velocity 42, 43
driving force 18
ductile materials 40

E

earth wire 51
efficiency 32, 33
elastic deformation 35, 40, 41
elastic limit 34, 35, 36, 37, 40
elastic potential energy 30
elastic strain energy 36, 37, 39
electric fields 62
electrical energy 48–50
electricity bills 50
electromagnetic spectrum 62, 63
electromagnetic waves 58, 60
electromotive force (e.m.f.) 52, 53
electron charge 42
electron diffraction 69, 80
electron microscope 81
electrons 42, 43, 46, 74, 76–78
electronvolts 74, 78
energy 28–33, 36–39, 48–50, 52, 54, 58, 62
energy levels 78, 79
energy transfers 32
error analysis 74, 75
extension 34, 36–38

F

filament lamps 46
fluid friction (drag) 18
force-extension graphs 36
forces 8, 18–20, 22–26, 28, 29, 34, 36–38
free fall 8, 9
frequency 58, 59, 62, 65–67, 70, 74, 76–78
friction 18, 19, 26
fringe patterns 69
fringe spacing 71
fundamental frequency 66
fuses 51

G

Galileo 10
gamma rays 58, 62, 63
global positioning systems (GPS) 27
gravitational field strength 17, 20
gravitational potential energy 28, 30, 31
gravity 6, 8, 9, 16, 17, 20, 21
ground state 78

H

harmonics 66
heat 18, 32, 33, 58, 62, 63
Hooke's law 34, 35, 37, 40

I

I/V characteristic graphs 46, 47
inclined plane experiment 10
inertia 20
infrared 62, 63
insulators 43
intensity 58, 61, 76, 77
interference 64, 65, 70–72, 80
internal resistance 52, 53
ionisation 58, 63
ions 43, 48

J

joules 29, 50

K

kilowatt-hours 50
kinetic energy 18, 28, 30, 31, 74, 76, 77
Kirchhoff's laws 54

L

laser light 68, 70
light 58, 60–63, 69–76, 78, 79
light-dependent resistors (LDRs) 47, 56
light-emitting diodes (LEDs) 74, 75
limit of proportionality 38, 40
line absorption spectra 79
line emission spectra 78, 79
lions 12, 14, 20
load (mechanical) 34, 35, 38, 41
load resistance 52, 53
load-extension graphs 34
longitudinal waves 60
lost volts 52
loudness 58

M

magnetic fields 62
magnitude 4
Malus' law 61
mass 4, 16, 17, 20
metallic conductors 46
microwaves 62, 63, 67, 70
Millennium bridge 67
moments 24, 25
momentum 80, 81

Index

N

Newton's laws of motion 16, 17
nodes 66

O

ohm (definition) 44
ohmic conductors 45, 46
Ohm's law 45

P

parachuting 19
parallel circuits 54
particle theory of light 74, 80, 81
path difference 64, 65
peer review 1
period 58, 59
phase 58, 64, 65
photoelectric effect 74, 76, 77, 80
photoelectrons 76
photon emission 78, 79
photon model 74, 76
photons 74, 76–80
pivots 24, 25
Planck's constant 74, 75, 77, 78, 80, 81
plastic deformation 35, 40, 41
polarisation 60–62
polarising filters 60, 61
polymeric materials (polymers) 41
polythene 41
potential difference 43–45, 46, 48, 49, 52
potential dividers 56, 57
potential energy 30, 31
potentiometers 57
power (electrical) 48–51
power (mechanical) 29
power rating (of an appliance) 51
power supplies (HT and EHT) 52
probability wave 80
projectile motion 10, 11

Q

quanta 74

R

radio waves 62, 63
raisin d'être 43
random error 74
rarefactions 60
reaction force 16, 28
reaction time 26
reflection 58, 60, 62
refraction 58, 62
resistance 44–49, 52–56
resistivity 44
resistors (combinations of) 54–56
resolving forces 4, 22, 23
resolving vectors 4, 5
resonance tubes 67
resonant frequency 66
resultant force 16, 18, 23
ripple tank 68
rotating knives 25
rubber 34, 41

S

safety cages 26
Sankey diagrams 32, 33
scalar quantities 4
scientific process 1, 2
seatbelts 26
semiconductors 43, 46
sensors 46, 56
series circuits 54
shock waves 60
sound waves 58, 60, 65–67
spectra 73, 78, 79
speed 4–7, 14, 29
speed (of a wave) 59
speed of light 63, 74
speed of sound 67
stability of objects 21
standing (stationary) waves 66, 67
stiffness 34
stiffness constant 30, 34, 37
stopping distance 26
strain (tensile) 36–41
stress (tensile) 36–41
stress-strain graphs 36, 39–41
stringed instruments 66
sunscreens 63
superposition 64–66
systematic error 74

T

temperature 45–47, 56
tensile forces 34, 36–38
terminal p.d. 52
terminal velocity 18, 19
thermistors 47, 56
thinking distance 26
threshold frequency 76, 77
threshold voltage 47, 74
time 4, 6–8, 12–15, 26, 27, 29, 42
torque 24, 25
transistors 56
transverse waves 60, 61
trilateration 27
two-source interference 70, 71

U

ultimate tensile stress 36
ultraviolet (UV) radiation 62, 63
uncertainty 74, 75
uniform acceleration 6, 7, 9–11
units of electricity 50

V

variable resistors 57
vectors 4, 5, 16, 18, 19, 22, 23, 42
velocity 4, 6, 9, 13, 29
velocity-time graphs 14, 15, 18, 19
viscosity 18
visible light 62, 63
volt (definition) 43
voltage 43–49, 51–57
voltmeters 53, 56, 57

W

water waves 68
watt (definition) 48
wave equation 59
wave theory (of light) 76
wave-packets 74
wave-particle duality 74, 80, 81
wavelength 58, 59, 62, 65–74, 78–81
waves 58–74, 76, 80, 81
weight 19, 20, 21
wind instruments 66
work 28, 29, 37
work function 76, 77

X

X-rays 62, 63

Y

yield point 40
Young Modulus 38, 39
Young's double-slit experiment 70